138045

SCIENCE FAIR PROJECTS
The Environment

Bob Bonnet & Dan Keen
Illustrated by Frances Zweifel

Sterling Publishing Co., Inc. New York

Edited by Claire Bazinet

Library of Congress Cataloging-in-Publication Data

Bonnet, Robert L.
    Science fair projects : the environment / by Bob Bonnet and Dan Keen;
illustrated by Frances Zweifel.
        p.   cm.
    Includes index.
    ISBN 0-8069-0542-5
    1. Science projects—Juvenile literature. 2. Science—Exhibitions—Juvenile
literature. [1. Science projects.]
I. Keen, Dan. II. Zweifel, Frances W.   III. Title.
Q182.3.B674   1995
507.8—dc20   94-46331
                                                                  CIP
                                                                  AC

10 9 8 7 6 5 4 3 2

Published by Sterling Publishing Company, Inc.
387 Park Avenue South, New York, N.Y. 10016
© 1995 by Bob Bonnet and Dan Keen
Distributed in Canada by Sterling Publishing
℅ Canadian Manda Group, One Atlantic Avenue, Suite 105
Toronto, Ontario, Canada M6K 3E7
Distributed in Great Britain and Europe by Cassell PLC
Wellington House, 125 Strand, London WC2R 0BB, England
Distributed in Australia by Capricorn Link (Australia) Pty Ltd.
P.O. Box 6651, Baulkham Hills, Business Centre, NSW 2153, Australia

Sterling ISBN 0-8069-0542-5

This book is
hereby dedicated:

To our wives,
Shannon Bonnet
and Rhonda Keen

CONTENTS

A Note to the Parent

Welcome to the exciting exploration of the world around us...the world of science. Our environment provides us with many things to observe and processes to understand. Knowledge is gained by observing and questioning.

Science should be enjoyable, interesting, and thought-provoking. That is the concept the writers wish to convey. While this book presents many scientific ideas and learning techniques that are valuable and useful, the approach is designed to entice the young child with the excitement and fun of scientific investigation.

The material is presented in a light and interesting fashion. The concept of measurement can be demonstrated by teaching precise measuring in inches or centimetres, or by having a child stretch his or her arms around a tree trunk and asking, "Are all children's reaches the same?" We present science in such a way, so that it does not seem like science.

The scientific concepts introduced here will help the young student to later understand more advanced scientific principles. Projects will develop those science skills needed in our ever-increasingly complex society: skills such as classifying objects, making measured observations, thinking clearly, and accurately recording data. Values are dealt with in a general way. One should never harm any living thing just for the sake of it. Respect for life should be fundamental. Disruption of natural processes should not occur thoughtlessly and unnecessarily. Interference with ecological systems should always be avoided.

The activities presented in this book target third- through fifth-grade students. The materials needed to do the activities are commonly found around the home or easily available at minimal cost.

Because safety is and must always be the first consideration, we recommend that all activities be done under adult supervision. Even seemingly harmless objects can become a hazard under certain circumstances. For example, a bowling ball can be a danger, if it is allowed to fall on a child's foot.

There are many benefits in store for a child who chooses to do a science project. It motivates the child to learn. Doing such an activity helps develop thinking skills: it prompts a child to question, and learn how to solve problems. In these activities, the child is asked to make observations using all the senses and record those observations accurately and honestly. Quantitative measurements of distance, size, and volume must be made. Students may find a subject so interesting that, after the project is completed, they will want to do more investigation on their own. Spin-off interests can develop. In doing a science project about weather, while using a computer to record weather data, a child may discover an interest in computers.

The authors recommend parents take an active interest in their child's science project. Besides the safety aspect, when a parent is involved, contact time between the parent and child increases. Such quality time strengthens relationships, as well as the child's self-esteem. Working on a project is an experience that can be shared. An involved parent is telling the child that he or she believes education is important. Parents should support the academic learning process at least as much as they support Little League, music lessons, or any other growth activity. Parents should help the student in reading, understanding, and completing these educational and fun projects.

Adults can be an invaluable resource for information that the child draws upon, while older people share their own life experiences. They can also provide transportation, taking the child to a library or other places for research. In our school, one student was doing a project on insects and his parents took him to the Mosquito Commission Laboratory, where he talked to professionals in the field.

The projects in this book have been designed as "around-you science," in contrast to book science. By "around-you science," we mean doing a science project right where you are—in your home, your neighborhood, your school. Investigations can even begin right at your feet. What is living under that old board lying on the ground? What species of insect are lying now on your windowsill, trapped by the screen? Are termites eating your house? How many rings can you count in the trunk of the tree your neighbor has just cut down? Are some rings closer together

than others? Why? Since the environment takes in everything around us, the projects in this book cross over into many science disciplines: biology, weather, physics, botany, chemistry, behavior, and even consumerism. Get excited with your child about the world around us!

Clear and creative thought is a primary goal for the young scientific mind. This book will help prepare a young person for future involvement and excelling in the field of science.

BOB BONNET AND DAN KEEN

Project 1

NOWHERE TO BE SEEN

Finding evidence of animal presence

The planet Earth is full of living things. Many times we can tell that an animal has been in a place, even though it is not there now. Hypothesize that you can prove that animals have been in your house, yard, neighborhood, or park. See if you can find proof that ten different animals have been there and gone. Before you start, think about what kind of animals might be around. Make up your "guess list," then start your search for evidence. Make a record of whatever you find.

Some things to look for are: animal hair on the furniture, a smell (a skunk is a good example), a sound (a woodpecker or cricket may be heard but not seen), a mound of sand or a hole in the ground, a chewed leaf, a nest, scratched trees or broken branches, a cocoon or a web, waste products, skeletons or empty shells, holes in trees, tracks on the ground. Look closely. What do you see?

You need
- an adult
- outdoor area
- pencil and paper

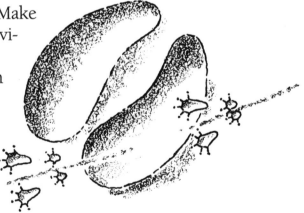

Something more

Can you find evidence of something that animals did? If you find parts of a fly in a spiderweb, even if you don't see the whole fly or the spider, you can guess that there was a struggle between these two animals.

11

Project 2

FRUIT FOR THOUGHT

Using senses to recognize fruits

We gather information about our environment (the things around us) by using our five senses: seeing, hearing, touching, smelling, and tasting. We use different senses to know about different things. You can't listen to garlic, or smell the sound of music.

To study the senses, let's test the ability of someone to identify fruits by using each sense of the five senses one at a time. Before that person comes, remove a small section of peel from each

of four fruit (orange, apple, grapefruit, and banana) and place one in each of the four containers. With a marker, number the lids (not see-through) 1, 2, 3, and 4. Make a list of the numbers and the fruit peel inside each container.

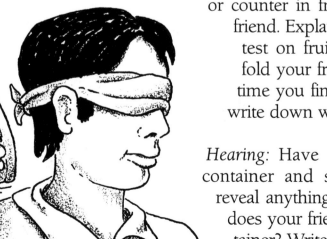

Put the four containers in a row on a table or counter in front of your test subject or friend. Explain that you are conducting a test on fruit identification, then blindfold your friend and start the test. Each time you finish testing one sense below, write down what your friend tells you.

Hearing: Have your friend pick up each container and shake it. Does the sound reveal anything about the fruit peel? What does your friend tell you about each container? Write it down.

Smelling: One at a time, have your friend open a container and smell the contents. Write down what your friend tells you.

Touching: This time, have your friend open each container and reach inside. Can you friend recognize a fruit by feeling the peel? Again, write down what your friend tells you.

Tasting: Now, again one at a time, let your friend taste each peel, or a section of the fruit. What does your friend say now?

Seeing: Take off the blindfold. Let your friend open each container, look inside, and tell you now, if he or she can, which fruits they are. Write it down.

How did your friend do? Do this activity with other friends. How many do you think will be able to guess the fruits? Which senses work best for this activity? Which senses did not help much?

Something more
Use other things in the test. Try smelling different odors, tasting different liquids. Can you feel the difference between coins?

Project 3
THE GREAT SOAP RACE

Comparing how long soaps last

Soap is one of the most common things found in the home. We use soap to clean our houses, hair, cars, dishes, and our bodies. The soaps we use to clean our bodies come in bars of different shapes, smells, and sizes. When we use a bar of soap to wash up, it gets smaller. Which brand of soap lasts the longest?

Go to the store with an adult and help select four different kinds of bar soap. Ask to have a small piece of soap cut from each bar. Each piece should be the same size.

You need
- an adult
- 4 small pieces of different brands of soap
- 4 pieces of paper
- 4 clear glasses of warm water
- pencil and paper

Get four small pieces of paper. On the first piece write #1 and the name of one of the soap brands. On the second piece of paper, write #2 and the name of another soap brand. On the third piece of paper write #3 and the name of the third brand of soap. On the fourth piece of paper write #4 and the name on that bar of soap.

Place the four pieces of paper next to each other on a table. Fill four clear drinking glasses with warm water. Each glass should contain the same amount of water. On each piece of paper, place one glass of warm water and put a piece of the soap brand written on the paper into the water.

After a while, look at the pieces of soap. Which one is the smallest? Which piece is the largest? Look again later. Did the smallest piece dissolve completely? How long did it take for the different brands to dissolve?

Don't waste, recycle. Put the soapy water into a plastic bottle to use as liquid soap.

Something more

Do you think that the soap that lasted the longest would make the most lather? Which brand costs the most? Does the longest-lasting soap also cost the least? Does the soap that lasts the longest also smell the best? If a bar of soap has a nice smell, does it last longer than most other soaps?

Do any of the soaps float?

Project 4

PEOPLE MIGRATION

Determining traffic patterns

Some animals migrate. That means they move from one place to another. If you live in a northern climate, there will be some birds that fly south for the winter. They come back in the spring when the weather gets warmer. In mountainous areas, some animals migrate to the valleys in the winter and go back up in the summer. They have trouble finding food in the mountains in the winter snow, so they come down where there isn't as much snow.

> **You need**
> • pencil and paper
> • a two-way street or road near your home
> • an adult, in high traffic areas

How do people move during the day? Do they drive one way in the morning to go to work and then travel the other way to go home? Stand on a sidewalk or roadside near your home and count the number of cars that pass by. Do this for fifteen minutes before you go to school in the morning, and then again for fifteen minutes in the early evening. A friend can help by counting the cars going one way while you count the cars going the other way.

To count the cars, draw one line for each car on a piece of paper. Do this for four cars, and then draw a line across the four lines to count the fifth car. This makes groups of five. It is easier to add numbers together if they are grouped. Do

CARS →
15 minutes in MORNING:
卌 卌 卌 II

15 minutes in AFTERNOON:
卌 卌 卌 卌 IIII

← CARS
15 minutes in MORNING:
卌 卌 卌 卌 卌 卌 卌 卌 II

15 minutes in AFTERNOON:
卌 卌 III

you think that many of the people who go by your house in the morning, go back the opposite way at night?

Something more

Look at the numbers from the morning and afternoon counts. When were there more cars? If more cars went one way in the morning, did more cars go the other way in the afternoon? If more cars go one way one morning, will this be true every morning? Try counting the passing cars at the same time for five days, beginning Monday and ending Friday. Are there always more cars going one way than the other in the morning?

Many people work Monday through Friday. How do the numbers change if you count on a Saturday or a Sunday?

You may want to try counting at different times of the morning and afternoon. You could count for more than fifteen minutes, perhaps a half hour or an hour.

Is there anything near your home that might affect your count? A cinema, sports arena, shopping mall, or school?

NUMBER OF CARS IN 30 MINUTES

Project 5

ORDER OUT OF CHAOS

How to classify things

One of the skills scientists must have is to be able to classify things. Things can be put into groups, or classifications, when something about them is the same, like color or size.

Gather twenty different stones from around your neighborhood. Put three paper plates on a table. Group the stones by their size. Put small stones on one plate, medium-sized stones on another plate, and the biggest stones on the last plate. Count how many stones you put in each group, and write the number down

Empty the plates and put all the stones back in one pile. Now group them by light and dark color. Put light-colored stones on one plate, dark ones on another plate, and mixed or medium ones on the last. Count how many stones you put in each group and write it down.

> **You need**
> - a shoe box
> - 20 different stones from your neighborhood
> - 3 paper plates
> - pencil and paper

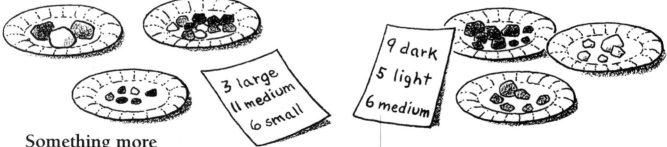

Something more

Can you think of other ways to group your stones? Can you group them by heavy and light? Can you group them by smooth and rough?

Find out which stone is the hardest by seeing which one will scratch another. Make a chart showing the characteristics of each stone. Pick up a geology book at your library to identify your stones and find out about them.

Project 6

HOME COMFORTABLE HOME

Temperature in your home environment

There's no place like home ... a place where you feel happy and comfortable. Even though your home has heat in the winter and fans or air-conditioning in the summer, some places in your home may not be as comfortable as others because of the temperature. Some spots may be comfortable at one time of the day, but uncomfortable at another time.

You need
- thermometer
- pencil and paper
- an adult (for safety around heat sources)

Measure the temperatures of different areas of your home in the morning, at noon, and in the evening. Which are the most comfortable areas, and at what times of day? Write down the location, the temperature, the time, and whether you are comfortable or uncomfortable there.

Sitting by a window may be comfortable and warm during the day, when the sun is shining in, but cold and uncomfortable in the evening, when a leaky window allows a cold winter draft to blow through.

What is the temperature near your home's source of heat (air ducts, radiators, electric baseboards, fireplace)? Is there a spot near a sunny window that doesn't need any other heat when the sun is shining in?

Something more
In a heated room, do you feel uncomfortable even though the temperature is just right? If your skin feels dry, maybe it's because there is not enough moisture in the air.

Project 7

NUKED BEANS

The effect of microwaves on seeds

Microwave ovens use powerful radio (electromagnetic) waves to heat food. These waves can be dangerous to people; that is why the makers of these ovens build them in a way that keeps the microwaves safely inside. They put special switches on the oven doors: the switch turns the power off if anyone opens the door. This keeps the people who use the microwave ovens safe from the invisible waves inside.

What about those unseen radio waves trapped inside the oven? Do you think that they could hurt seeds?

Fill an egg carton with potting

You need
- an adult
- microwave oven
- 12 bean seeds
- potting soil
- plastic or plastic-foam egg carton
- water
- marking pen
- pencil and paper

soil. Use a carton that is made of plastic or plastic foam so that it won't leak when water is poured into it.

Use a marking pen to write the number "0" by the

first two pockets, "5" by the second set of pockets, "10" by the third, "15", "20", and "25" by the next ones. These numbers stand for the number of seconds the beans will spend in the microwave.

Put a bean in each of the two pockets by the number "0". The zero means that these seeds spent no time in the oven.

Have an adult put two beans in a microwave oven and turn it on for 5 seconds on a low power setting. Then plant the beans in the egg carton in the pockets marked "5". Have an adult put two more beans in the oven for 10 seconds. Plant them in the "10"-second egg carton pockets. Do the same thing for 15, 20, and 25 seconds. Plant the beans in the right holes.

We are planting two seeds instead of just one to give us a true test. Not all seeds "germinate," or begin to grow. Sometimes there may be something wrong with a seed and it won't germinate. By planting more than one seed each time, there is a better chance of a successful test.

Pour water onto the beans. Be sure they are thoroughly wet, but not swimming in water. Close the lid of the egg carton to keep the seeds moist. Check them each day and add water equally to all the seeds if they look dry. Write down what you see each day. What do you think will happen? Hypothesize that the seeds that were in the oven the longest will not germinate. (A hypothesis is one possible answer, or guess, based on what you know. The hypothesis must be tried and tested to see if it is true.)

After two weeks, look at your results and see if your hypothesis was right.

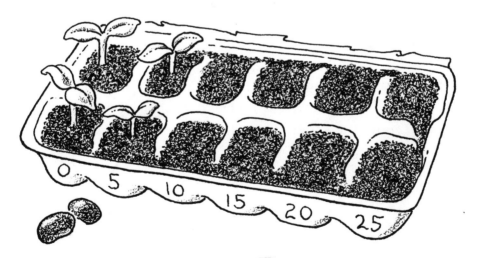

Something more
Try other fast-sprouting seeds, such as peas or radish seeds. Do the test using bulb plants.

Project 8

FADE OUT

How sunlight affects colors

A sunny window is a great place to put a plant, but it might not be a good place to put your couch. Plants do well in sunlight, but how do couches do? What if sunlight does something to the color of cloth. Hypothesize that sunlight can fade the color of things.

Gather two pieces of red, orange, yellow, green, blue, and black construction paper. Cut two strips of red paper about 2 inches (5 cm) wide and 6 inches (15 cm) long. Cut two pieces each of orange, yellow, green, blue, and black the same length.

Using tape, stick one of each of the colored strips onto a window that gets a lot of sunlight. Take the other colored strips and place them in a dresser drawer, away from sunlight.

After two weeks, take the colored strips off the window. Take the other strips out of

You need
- a sunny window
- colored construction paper
- adhesive tape
- safety scissors
- dresser drawer

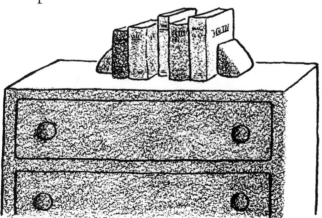

the dresser drawer. Set the strips next to each other, matching them by color. Can you see any difference between the colored strips that were kept in the drawer and the ones that were placed in the sunny window? Did any of the colors fade? Which color faded the most? Which color faded the least? If you put a couch in a sunny window, which color cloth do you think would be changed the least?

What would happen if the strips were left in the window for a month?

Something more

Gather two pieces of red, orange, yellow, green, blue, and black construction paper. Cut two strips of red paper about 2 inches (5 cm) wide and 6 inches (15 cm) long. Cut two pieces of orange, yellow, green, and blue the same length. With a black piece of construction paper, cut 5 strips measuring 2 inches wide (5 cm) by 3 inches (8 cm) long. Take a black strip and tape it over half of the red strip. In the same way, tape a black piece of paper over half of the orange, yellow, green, and blue strips. Half of each colored strip will be visible and half will be covered by black paper. Using tape, stick these strips on a window that gets a lot of sun. The side of the strips that have the black paper on them should be facing outside. Take the other colored strips and place them in a dresser drawer, away from sunlight.

After two weeks, take the colored strips off the window. Take the strips out of the dresser drawer. Set the strips next to each other, matching them by color. Take the black strips off. Can you see any difference in the colored strips that were kept in the drawer and the ones that were placed in the sunny window? Is there any difference between the parts of the colored strips that were covered by the black paper and the colored paper kept in the drawer? Did any colors fade even though they were covered by black paper?

How do other colors, like brown, purple, and white, do in sunlight?

Project 9

BLOWING IN THE WIND

Comparing evaporation indoors and out

All plants must have a certain amount of water to grow. Garden plants, that live outdoors, may need to be watered more often than plants that are always indoors. A garden may need water more often than potted plants. The sun and the air outside dry the soil. Hypothesize that soil dries more quickly outdoors than indoors.

Check the weather forecast. The next two days will need to be sunny.

Fill two empty one-pound plastic butter tubs with potting soil. Leave the soil loose in the tub. Do not pack it down. Use a small food scale or a balance beam to see that the containers weigh the same (You can make a simple balance beam by laying a ruler across a pencil,

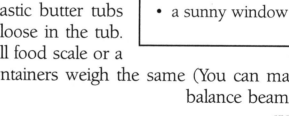

You need
- 2 one-pound plastic butter tubs
- potting soil
- measuring cup
- food scale
- 2 sunny days
- water
- a sunny window

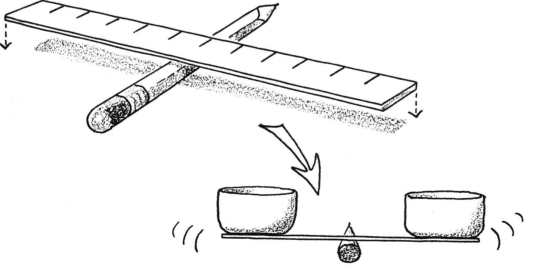

or hanging a coat hanger from a string and then using a paper clip at each end to hold light materials.) If one container is heavier, scoop a little of the potting soil out with a spoon until the tubs and their soil weigh the same.

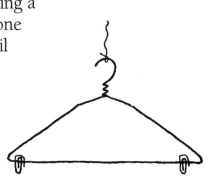

Fill a kitchen measuring cup with water to ½ cup. Pour all of the water into one tub. Fill the measuring cup again to ½ cup. Pour all of the water into the other tub.

Put one tub outside your house in the sun. On the same side of the house, put the other tub inside in a sunny window.

At the end of two days, weigh each of the tubs on the kitchen scale or compare them using a balance beam. The lighter tub will have had more water dry out of it.

Something more
Is there any difference if the days are cloudy? Does it make a difference if the wind is calm or strong?

Project 10

SHAKE, RATTLE, & GERMINATE

Breaking down seed coats

Many seeds have a coat covering them. Often, before a seed can germinate and start to grow, its hard coat needs to be weakened or broken down. Hypothesize that seeds whose coats have been weakened by scratching will start to grow before seeds that are not scratched.

Ask an adult to cut out two round pieces of rough sandpaper using old

You need
- a one-pound plastic butter tub
- sandpaper (80 grit)
- old scissors
- an adult, to cut sandpaper
- bean seeds
- watch or clock with a second hand
- potting soil
- 4 egg cartons
- glue
- marking pen
- water
- pencil and paper

scissors. Glue one piece to the inside bottom of a plastic butter tub or margarine container, and the other piece inside the tub's lid.

Next, fill six pockets in each of four egg cartons with potting soil. Put six beans in the soil in one egg carton. Each seed should be planted in its own separate pocket. Use a marking pen to write "Not shaken" on the top of the egg carton.

Place six other bean seeds inside the tub with the sandpaper. Shake the tub as hard as you can for ten seconds. Put the seeds in an egg carton. Write "Shaken 10 seconds" on the lid of the carton.

Put six bean seeds inside the tub and shake as hard as you can for 30 seconds. Plant these seeds in another egg carton and write "Shaken 30 seconds" on the lid. Do the same thing for 60 seconds. Write the time on the lid of the egg carton.

Sprinkle water on all of the seeds. Close the lids to keep them from drying out.

Every day, open the lids and see if any seeds have begun to grow. Write down on the paper the date and anything you observe. If the potting soil feels dry, water all the seeds equally.

After one or two weeks, look at your notes. Which seeds began to grow first? Did the shaking cause any of the seeds not to grow at all?

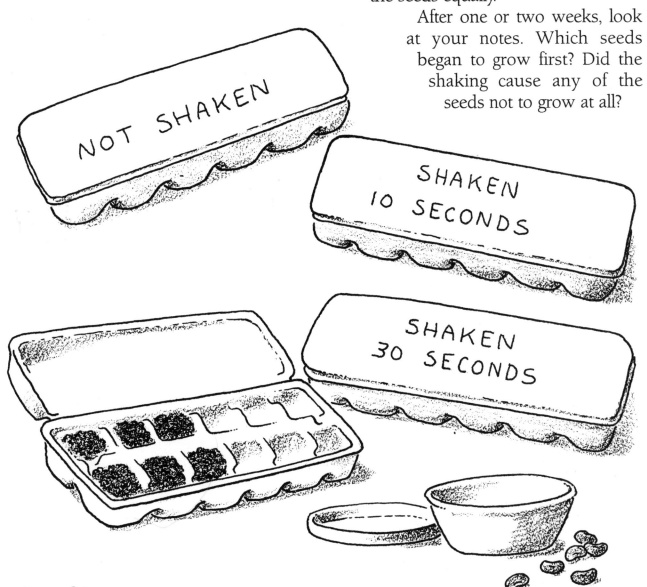

Something more

Will the results be the same for all seeds? Try morning glory, radish, and other kinds of flower or vegetable seeds.

Does soaking in water, tea, or other liquid affect how fast seeds germinate? Soak some seeds for one day, soak others for three days, and others for five days.

Project 11

YOU BUG ME

Attracting insects to sweetened liquids

All of us have seen flying insects gather around a porch light at night during the warm and hot times of the year. We know that many insects are attracted to light. Are there other things that attract them? Do they like sweet things? Hypothesize that they do.

Get four small cereal bowls: they must all be the same size. Pour soda into one bowl. Pour milk into another bowl. Pour water into the third bowl. Pour pancake syrup into the last bowl.

You need
- porch with a light
- 4 cereal bowls
- picnic table
- pancake syrup
- soda
- milk
- a warm night

When night comes, put the four bowls on a table outdoors, underneath a porch light. The bright light will attract insects to the table. After one hour, look at the four bowls. Count the number of insects you see in and around the edge of each bowl. Are there more insects at the sweet pancake syrup and soda bowls than there are at the milk and water bowls?

Something more
Get a book on insects. Can you identify any of the insects that visited your bowls?

Project 12

JUST A DRINK OF WATER

Moisture needed to germinate seeds

A seed needs water to germinate, but how much water does a bean seed need? Hypothesize that if a seed does not get enough water, it will not begin to grow. This often happens in nature, if there is not enough rain or the air is too dry.

With a marking pen, write the numbers 1 through 12 on an egg carton, putting one number near each pocket that normally holds an egg. Fill all 12 pockets with potting soil. Put one bean seed in each pocket.

Fill an eyedropper with water. Every morning at the same time, such as 8 o'clock, put one drop

You need
- eyedropper
- bean seeds
- water
- plastic or plastic-foam egg carton
- potting soil
- marking pen
- pencil and paper

of water on the seed marked "1", two drops of water on the seed marked "2", three drops on the seed marked "3", and so on. Always leave the lid of the egg carton open. Do this again twelve hours later, at 8 o'clock in the evening. Water the beans in this way every day for two weeks. Look at the project every day and write down what happens. At the end of the two weeks, see which seeds germinated. Did the seeds that were given only a few drops of water a day germinate?

Something more
Do the same project again, but this time use two egg cartons with seeds. Keep the lids closed on them. (Closing the lid should help keep in warmth and moisture.) Put one carton of seeds in a sunny window and the other in a dark place.

Project 13

BANANA COOKER

Ripening fruit by heat and light

People sometimes put fruit and vegetables in a sunny window to make them ripen faster. Is it the light from the sun or its warmth that causes them to ripen? Choose which you think is right.

Take two shoe boxes Paint the outside of one box with black paint. Paint the outside of the other shoe box white. Let the boxes dry.

Place the two shoe boxes in a sunny window. Put one green banana in each box. Place a thermometer next to each banana. Put the lids on.

Put the last banana on the window‚sill with a thermometer next to it.

Look at the temperature on the thermometers four times a day: when you get up, at lunchtime, before dinner, and at bedtime. Pick them up by the tube to read them. Be careful not to touch the metal tip or you will change the tempera-

<table>
<tr><td>

You need
- 3 green (unripe) bananas
- 3 thermometers
- 2 shoe boxes
- black paint
- white paint
- paintbrush
- a sunny window
- pencil and paper

</td></tr>
</table>

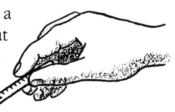

ture. Write down the readings.

Do this every day for two or three days. Look at the bananas. Are all the bananas turning yellow at the same time? Touch them. Is one softer than the others? Do you think softer means riper?

Something more

Do some of the yellow bananas have many black spots on the skin? Do the black spots mean that the banana has gone bad? Peel the three bananas and taste and compare them Are they different?

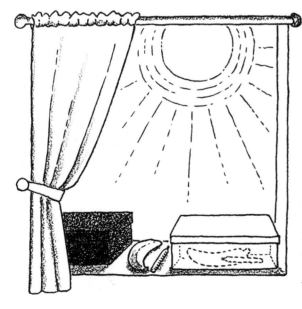

Project 14

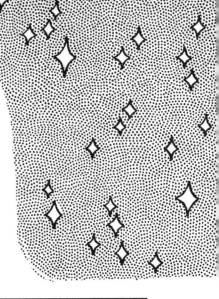

NIGHT SKY

A matter of seeing stars

Before the invention of the telescope in the early 1600s, the naked eye was the only tool people had to study and learn about the night sky. How many stars can you see, without using a telescope or binoculars?

On a nice night, without clouds or a bright moon, take blankets or sleeping bags, large containers (one per friend), and beans, and go outside. The area should be as dark as possible, so turn off any outdoor lights you can. Wait ten minutes or so for your eyes to adjust to the darkness, then lie on your back and look up at the stars.

First, decide where you are going to begin counting. You do not want to take your eyes away from the sky while you are counting or you will lose your place. Then, drop a bean into a bowl for each star you see. When you have counted all the stars you can see, take out the beans and count them.

How many stars did you count? Write it down. How many stars did each of your friends count? Have another star-count outing on a night when the moon is full. Compare how many stars you can count.

> **You need**
> - a clear, dark night (no bright moon or lights)
> - some friends
> - large bowls or jars
> - large bag of beans
> - pencil and paper

Something more

How do the bright lights from neighbors' houses, or in small towns or big cities, affect the number of stars you can see? Do you think light pollution is a problem for astronomers? What can they do about it? Read a book about the large telescopes and find out where they are located. Maybe you can visit one.

After being outside in the dark for a while, go into a bright room in the house. After a few minutes, go back outside. How long does it take for your eyes to adjust to the dark again?

Project 15

BACKYARD FORECASTER

Predicting weather by wind direction

Some people use a barometer, an instrument that measures air pressure, to help forecast the weather. When the barometer's needle is rising, fair weather is usually coming. When the needle falls, rain is likely.

What about wind direction? Hypothesize that wind direction can also be used to indicate fair or stormy weather, then test it out.

You need
- a long thin strip of light-weight cloth
- a wooden stake about 4 feet (120 cm) long
- hammer
- magnetic compass
- thumbtack
- a week or two
- outside area
- pencil and paper
- bright-colored yarn
- 4 ice-cream sticks

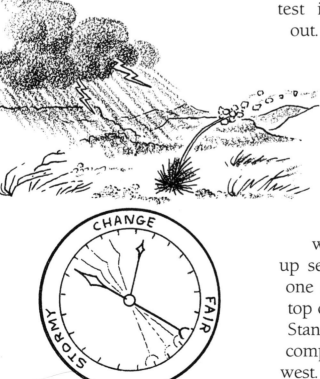

Find an area in your yard or other place nearby where there is nothing to block the wind. Hammer the wooden stake a little way into the ground, until it stands up securely. Using a thumbtack, fasten one end of a thin strip of cloth to the top of the stake.

Standing at the stake, use a magnetic compass to find north, south, east, and west. Mark a direction—N, S, E, and W—

at the tip of each ice-cream stick, then push them into the ground about three feet (90 cm) from the stake in each of the four directions. Tie a piece of bright-colored yarn around the stake and out to the ice-cream stick to the north (N). Do the same for the other three directions. This will help you see the wind direction from a distance.

Every day, once in the morning and again in the evening, look to see which direction the wind is

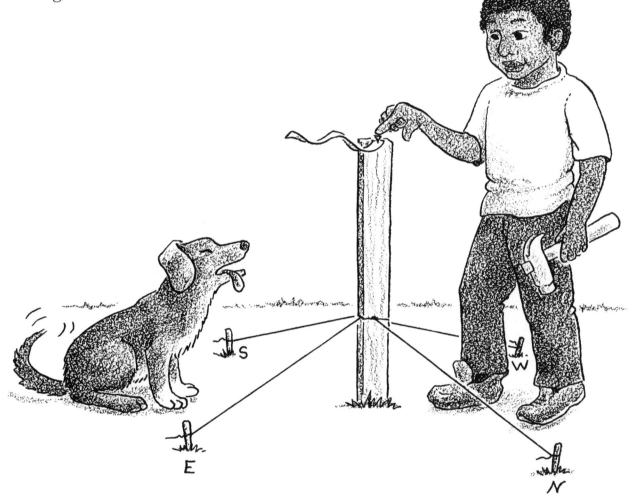

blowing the cloth. You need to record the wind direction on both fair days and stormy days. After a few weeks, examine your log to see the wind direction you recorded just before a storm and just before fair weather. Was your hypothesis correct?

Something more
Can you use your ribbon device to measure wind speed? Make a scale showing wind speed. The ribbon is lifted farther away from the stick when the wind is blowing harder.

Project 16

TUNNELLERS

The digging habits of ants

You can always tell where there is an ant nest underground, by the telltale mound of soil around a small hole. In nature, many things can block an ant hole. An animal might walk over it and push soil into the hole. A tree limb may fall during a storm and cover the entrance to the ants' nest.

> **You need**
> • 5 ant hills
> • pencil and paper

What do you think ants will do when their hole is covered? Will they dig the same hole out again, or will they make a new hole near the old one? Hypothesize which you think will happen.

Find five ant hills. Put a handful of soil over each hole. Do not harm any of the ants. Keep checking around the holes. Write down anything you see the ants doing. If the ants do the same thing at each hole, can you guess where the ants would dig if you covered another hole?

What if a very small twig or piece of straw were placed in the opening of an ant hole? Would the ants try to move it? Could they? What does that tell you about ants?

Something more
Do you think that ants have more than one way of getting in and out of their underground homes? Build an ant farm, or buy one; then watch to see if the ants make more than one opening into their home.

Project 17

CLIP IN A BOTTLE

Comparing force fields

We are all familiar with the force of gravity and magnetic force. When a magnet is brought near something metal, the magnet tries to pull the metal object to it. When you try to jump up, the force of gravity pulls you back down towards the Earth. The force of gravity and magnetic force are invisible forces, but they are very strong. Can a magnetic force be stronger than gravity?

You need
- an adult
- a hammer
- safety pin
- clear glass jar with lid
- strong magnet
- thread
- adhesive tape

Have an adult poke a small hole in the lid of a clear glass jar, such as a mayonnaise jar or a jar that held apple sauce.

Turn the jar upside down and use adhesive tape to stick a magnet onto the bottom of the jar. The magnet should be on the outside of the jar.

Tie one end of a piece of thread to the small end of a safety pin. Push the other end up through the bottom of the lid. Screw the lid onto the jar. Slowly pull the thread up through the hole until the safety pin is standing straight up, with its tip just touching the bottom of the jar. Put a piece of adhesive tape over the hole in the lid to keep the hanging safety pin in place.

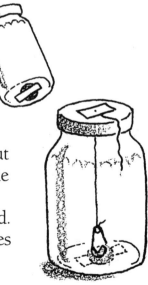

Slowly turn the jar over and place it upside down on its lid. Does the magnetic force keep the safety pin in place, or does gravity make it fall down?

Something more

Will a stronger magnet allow you to raise the safety pin off of the bottom of the jar? How high off the bottom of the jar can you raise it and still have the magnetic force strong enough to overcome gravity?

Project 18
EARLY RISERS
Animal sounds, day and night

Some animals sleep during the day and are awake at night. Others sleep at night and are active during the day. Animal sounds can be heard outside at any time: owls, birds, raccoons, frogs, crickets, locusts, cats, dogs, horses, cows, ducks. Hypothesize that the animal sounds that are heard in the morning are not the same as the ones heard after dark.

Early in the morning, after the sun has risen, go outside, announce the time of day for the tape recorder and then record about five minutes of animal sounds. In the evening, after dark, announce the time again and then record another five minutes of animal sounds.

Listen to the recordings you made. Are the animal sounds recorded in the morning the same as the ones heard at night? Make a list of the animal sounds you recorded.

> **You need**
> - outside
> - tape recorder
> - pencil and paper

Something more
Is there any time during the day when fewer animal sounds are heard? At what time of the day are the sounds the loudest?

What animals do you hear at certain times of the year but not at other times?

Project 19
COOL SMELLS
The effect of temperature on odors

A molecule is a very tiny particle of matter made up of groups of even tinier particles called atoms. Molecules move faster when they are heated than when they are cold. As the temperature increases, more molecules escape and travel into the air. This means that perfume molecules can be smelled more easily when the perfume is warmed, as when you put some on your skin.

Since molecules move more slowly when they are cold, hypothesize that perfume will not smell as strong if it is cold.

Tear a sheet of paper towel into three strips. Fold each strip three or four times, to make it thicker. Put three drops of perfume on each piece. Leave one section of towel on the table, at room temperature. Place another towel section in the refrigerator, and the last piece in the freezer. After several hours, use your nose to test which of the three perfumed pieces of towelling has the strongest smell.

You need
- paper towel
- perfume
- use of a refrigerator
- use of a freezer

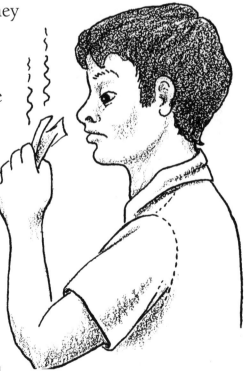

Something more
What do you think would happen if you were to put the perfumed towelling from the freezer into a sunny window?

Would molecules of garlic juice, lemon juice, and other "smelly" things react in the same way to heat and cold?

Project 20

BOXED-IN

A characteristic of recycled cardboard

Because they are concerned about the environment, many companies are using recycled cardboard and paper to make packaging for their products. Some boxes with the recycled symbol on them are light in color. Hypothesize that recycled "paperboard" is generally lighter in color than non-recycled cardboard.

Gather many different kinds of food boxes, such as contain noodles, sugar, gelatin, pudding, pretzels, and pet food. On a piece of paper, write the name of the product. Note whether or not the box is made of recycled paperboard (it will be marked on the box if it is), and if the color of the

You need
- a variety of food boxes (cereal, noodles, sugar, pudding,crackers, etc.)
- pencil and paper

box is light or dark. Since the outside of the package will likely have printing on it, open the box and look on the inside, where there is no printing.

Go over your notes. Did you find that all or most of the packages made from recycled paperboard were lighter in color? Did you find some boxes that didn't say they were recycled also had cardboard that was light in color?

SPAGHETTI - dark
CAT FOOD - light
CORNFLAKES - dark
SUGAR - light
SODA -
COCOA -
RICE MIX -

Something more

Research what "recycled" means in making packaging. Is .it made of all recycled material, or is there only *some* recycled material in it?

Write to a company that does not use recycled material in their packaging, and ask them to do it.

Project 21
ROCK COLLECTION, CITY-STYLE

Grouping rocks by their characteristics

Rocks are just about everywhere. If you live in the city or on the plains, you may still find a lot of different types of rocks in your community because people have brought them into your neighborhood. Rocks may be brought in from outside your neighborhood to make driveways, rock gardens, home decorations, landscaping, drainage, monuments, and building structures (like granite for stairs).

You need
- 20 different types of rocks from your neighborhood
- library books on rocks and minerals
- pencil and paper

Gather 20 or more different types of rocks from your neighborhood. Identify the rocks by using a book from the library, asking a science teacher, or asking anyone who knows about rocks.

Group them by hardness. Remember, you can determine if one rock is harder than another by trying to scratch one rock with another. The one that makes the scratch is harder than the one that gets scratched.

Something more

Are there any other characteristics about the rocks you have collected that are interesting? Are some rocks smooth? Why might that be?

Project 22

ARE ALL CANS CREATED EQUAL?

Comparing soda containers

Years ago, soda came in very thick and heavy glass bottles. Today, glass and other packaging materials are costly and we don't want to waste them, so soda comes in thinner glass bottles, plastic bottles, and thin aluminum cans that are recycled.

Do the makers of aluminum soda cans all use the same amount of aluminum in their products, or are some cans heavier than others? Do the manufacturers of 2-litre soda bottles all use the same quantity of plastic in their bottles?

Form a hypothesis, then collect 10 different brands of aluminum soda cans (leave the flip-tops on). Rinse out the cans and let them dry completely. Weigh each can on a scale and record the results. (If you don't have a gram-weight or postal/food scale, place two thumbtacks side by side, with the points up. Lay a ruler across them, so that the ruler balances like a seesaw. Place two cans, in turn, one on each end of the ruler, and see which is heavier).

Which of the cans is the heaviest? Which one is the lightest? If you have a scale, subtract the weight of the lightest can from the weight of the heaviest. Is there a big difference, or very little difference?

> ### You need
> - 10 same-size aluminum soda cans (12 oz.), all different brands
> - 10 2-litre soda bottles, all different brands
> - scale or homemade balance
> - pencil and paper

Something more
Do the same thing with plastic soda bottles. Is there a bigger difference between the weights of the bottles or the weights of the cans?

Project 23

A BETTER BATTER?

Why some foods hold heat

Some foods cool very quickly after being cooked or heated. Toast cools very fast, but a potato stays hot for a long time. Why?

If thicker foods hold their warmth longer than thinner ones, a thicker pancake should keep its heat longer than a thin one. Ask an adult to help you do this recipe experiment.

Mix two bowls of pancake batter. Make one batter thick. Add more milk to make the second batter thinner. Then, into a frying pan or on a grill, measure the same amount of each batter. Make one pancake using thick batter and the other using thin batter.

You need
- an adult, to make pancakes
- stove or grill
- utensils (spoon, knife, mixing bowls, frying pan)
- pancake mix
- milk
- stick of margarine
- two dinner plates

When both pancakes are all cooked, put each one on a plate. Cut two same size, ¼-inch slices from a stick of margarine that has been left at room temperature. Put one slice in the middle of each pancake. Watch the margarine. Does the slice pat on the thicker pancake melt faster than the one on the thin pancake?

Something more

Which cooked foods cool fastest after being served? Which foods stay hot the longest? Can anything be done to help keep foods warm longer? Will warming the plate before putting a pancake on it help to keep the pancake warm longer? Should syrup, honey, or jelly be warmed before putting it onto a pancake?

Project 24

TAMPER CHECK

Consumer safety awareness

Because something you eat or drink can be bad for you, food manufacturers list every ingredient on their packaging. All the ingredients have been tested, and are safe for most people to eat.

To keep people from opening up and tampering with foods before they are brought home, many companies make their packaging tamper-resistant. This makes the products safer to use. It also sometimes helps to keep foods fresher, so that they last longer.

The next time groceries are brought into your home, help to unpack them and check each item to see if it has any special packaging to help make it hard to tamper with. Peanut butter jars often have a foil seal under the lid. (There should be no pin holes or other breaks in the seal.) A vitamin bottle may have an extra plastic collar around the screw-off cap, in addition to the foil seal. The collar of a soda-bottle cap is made to separate when the bottle is opened. (Soda bottles also make a loud hissing or fizzing sound when they are opened for the first time.) Cans that are vacuum-packed for freshness, like peanuts, might make a "popping" sound when they are first opened. Jellies and other products vacuum-packed in glass jars often have bubble-top caps. The middle of the cap pops up when the vacuum is broken, so it is easy to see if the jar has been opened.

Look, listen, feel, and examine each product. Make a chart and write down the name of the product and how it is made tamper-resistant.

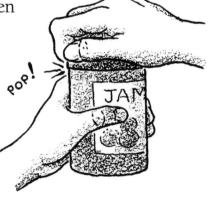

POP!

Something more
What should you do if you find something that might have been tampered with?

> **You need**
> - a variety of packaged goods from the grocery store (peanut butter, vitamins, bottled soda, etc.)
> - pencil and paper

Project 25

DYEING TO STAIN

Coloring with natural dyes

Dyes are used to color things. We can get many dyes from nature.

Get four small plastic butter or margarine tubs. Pour purple grape juice in one tub, cherry juice in another, and cranberry juice in the third tub. In the fourth tub, put in a teabag and pour warm water over it. Let it sit for 15 minutes.

Cut strips of cloth out of an old tee shirt or bed sheet. Dip one piece of cloth in each tub of natural dye. Use clothes pins to hang the strips on a clothesline to dry. Be careful handling them. You wouldn't want to stain your clothes.

Use your colored strips when you work on arts and crafts projects.

Can you find other things in nature that can be used as dyes? Try the purple berries of poke weed or other plants you have in your neighborhood.

You need
- pieces of old cloth
- teabag
- cranberry juice
- cherry juice
- purple grape juice
- 4 small plastic butter or margarine tubs
- warm water
- clothesline and clothespins

Something more
What cultures used natural dyes? In some cultures, people paint their faces. What do they use for paint?

Project 26

SPLISH-SPLASH IT OUT!

Understanding soil erosion

Have you ever started to water a flower bed but forgot to put the sprinkler on the end of the garden hose? If the water was turned on full force, soil probably splashed out of the flower bed, leaving big holes. That is why sprinkling cans and sprinkler hose attachments, with many small holes, are used to water pots of flowers and outside gardens.

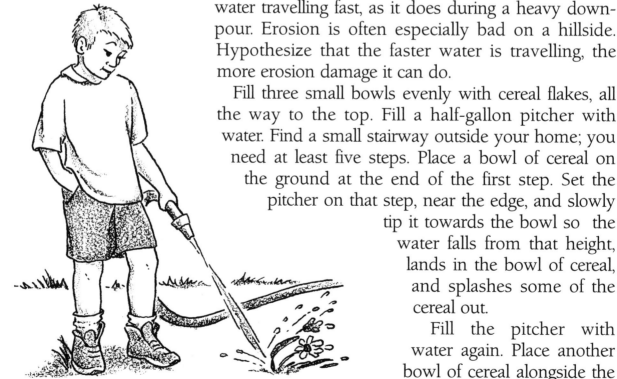

Erosion, or the wearing away of soil, is caused by water travelling fast, as it does during a heavy downpour. Erosion is often especially bad on a hillside. Hypothesize that the faster water is travelling, the more erosion damage it can do.

Fill three small bowls evenly with cereal flakes, all the way to the top. Fill a half-gallon pitcher with water. Find a small stairway outside your home; you need at least five steps. Place a bowl of cereal on the ground at the end of the first step. Set the pitcher on that step, near the edge, and slowly tip it towards the bowl so the water falls from that height, lands in the bowl of cereal, and splashes some of the cereal out.

Fill the pitcher with water again. Place another bowl of cereal alongside the

44

third step on the staircase. Set the pitcher on the edge of the third step and slowly tip it so that the water will again fall into the bowl.

Do this one more time, only put the pitcher on the fifth step.

Gravity makes things fall faster. The farther something falls, the faster it goes. Water poured from the fifth step will fall faster than water poured from the first step when it hits the bowl.

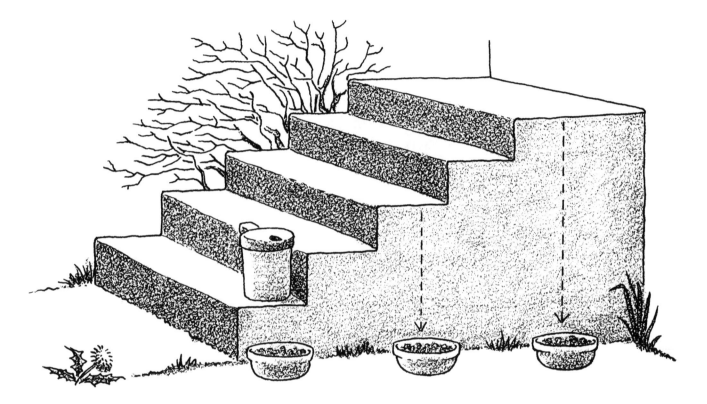

Look at each of the cereal bowls. In which one did the most "erosion" (loss of cereal) take place?

If your stairway has more than five steps, you can try pouring water from an even greater height.

If you wish, you can quantify, or measure, the amount of cereal that splashes out at each height by catching it in a tray or cookie sheet. Then let the cereal dry out and weigh it.

Leave the cereal used in this project as a treat for your neighborhood birds, squirrels, and chipmunks.

Something more
Put soil into a bowl or tray and place it under a downspout. What happens to the soil during a rainstorm? Try other kinds of soil. Is there a difference in the way it erodes?

Project 27

WEAR, WHERE?

Finding evidence of friction

Erosion is when soil wears away, usually caused by wind or moving water. Things can be worn away by friction, too. When two materials rub against each other, the friction from the movement can wear away the materials.

Look around your home, school, and neighborhood to find evidence of wear. Make a chart and list the things that you find that show wear. Write down where you find the wear, and make a guess as to what caused it. For example, at the public library, look at the steps. Do you see any difference on the part of the steps where people normally walk? Would you guess that the wear on the steps is caused by the friction of people's shoes on the steps? Look for wear on hand railings, floors, carpeting, doorknobs in public buildings, car tires, the end of a pencil, and the bottom of your shoes. Look at the ground around your school. Is there a dirt path where grass has been worn away by foot traffic? Look at the ground near a swinging gate. Is there an arc worn into the ground, where the gate swings back and forth?

You need
- your neighborhood
- pencil and paper

Something more

Do a "wear" report and share it with your classmates.

Project 28

DRIED OUT

Surface area and evaporation

Water evaporates, or disappears, into the air. The wet dew that collects on the grass in the morning usually evaporates by noon on a sunny day. How long would a cup of water take to evaporate? That depends on such things as the movement of the air and its dryness and temperature. But, those things being the same, the amount of the water's surface exposed to the air, called the "surface area," makes a big difference in evaporation time. If a cup of water is poured into a pie plate, where it can spread out, it will evaporate much faster than if the same amount of water is poured into a soda bottle that has only a small opening at the top.

Using a measuring cup, pour one-half cup of water into a soda bottle and one-half cup of water into a pie plate. Place them on a table by a sunny window. Several times a day, look at each one. Which is the first to become completely dry?

> **You need**
> • measuring cup
> • soda bottle
> • pie plate
> • water
> • a sunny window

Something more

Using three pie plates, set up an experiment to discover which evaporates faster: water, soda, or juice?

47

Project 29

COLOR ME HOT

Changing solar heating with color

The sun's light is energy that makes heat. When sunlight hits a dark object, much of the light is trapped. This makes the object warmer. When sunlight hits something light it bounces off (is reflected) and very little of the heat is trapped.

Fill the five drinking glasses with water. Each glass should be filled to the same level.

Choose five different-colored pieces of construction paper. One piece should be white, and one piece black. The other three pieces can be any other solid colors.

Wrap some black construction paper around one glass of water. Tape it in place. Then tape a piece of white construction paper around another glass and tape it. Wrap a different-colored paper around each of the other three glasses.

Put the wrapped glasses of water in a sunny window After about an hour of sun, put a thermometer in the first glass of water. Wait about three minutes, then read the temperature. Write it down. Measure the water temperature in each of the other glasses and write that down, too.

Is the water in the glasses covered by the dark paper warmer? Which is the warmest? Which glass has the water that is coolest?

You need
- 5 straight-sided glasses (same size)
- water
- construction paper (white, black, and others)
- adhesive tape
- a sunny window
- thermometer

Something more
Will a dark-colored liquid heat faster than a light-colored liquid? Use room-temperature milk and coffee, or add food coloring to water.

Project 30

RAINBOW FRUIT

Decorating with food

Color is very important in our lives. Colors are everywhere. Brightly colored packaging is used by manufacturers to make us want to buy their products. Even our emotions and moods can be changed by colors.

People use certain colors for holiday decorations. Black and orange are for Halloween. Green and red are Christmas colors. Light blue and light pink are for Easter. Can fruit be colored to make it fit the holiday?

Put a slice of banana, apple, peach, pineapple, and kiwi on a plate. Place a few drops of food coloring on each piece of fruit. Are any of the pieces of fruit easy to change their color? Try to color slices of other kinds of fresh fruits. Can you make a decorative plate of fruit slices for the next holiday?

You need
- slice of banana
- slice of apple
- slice of peach
- slice of pineapple
- slice of kiwifruit
- food coloring
- dinner plate

Something more

Can vegetables be colored with food coloring, too? Can you make a blue carrot or a green potato?

Write "Happy Birthday" on a banana, by slicing it along its length (the long way) and dipping a toothpick in food coloring to use as a pen. The toothpick will etch the letters into the banana.

The writing can be done either in dots, in lines, or both.

Project 31

LENDING LIBRARY

Recycling and extending the life of toys

If something can be used more than only once before throwing it away, it will help to reduce the amount of trash taken to landfills. Have your teacher set up a shelf in the classroom as a "lending library." Classmates can bring in books and games to lend to others. (Be sure the name of the person it belongs to is marked on each item, so that it can be returned.)

Some good items for the lending library are video-game cartridges (Nintendo®, Atari®, and others), books, comics, magazines, computer games, and videotapes. Classmates should ask their parents first if the item can be brought in to the lending library.

You need
• items collected from friends
• pencil and paper

Passing along to others things that we don't use anymore will help the environment by cutting down on what is thrown away. Extending the life of what we have will cut down on the number of new things we need to buy, and therefore need to manufacture, saving the Earth's resources.

Something more
Extend the life of things by repairing them. Start a toy repair shop in your classroom, where broken trucks and torn teddy bears can be fixed. Outgrown or unwanted toys can be donated.

Project 32

FIZZ MYSTERY

The secret that is soda

Did you ever taste "flat" soda? Soda that has lost its fizz, does not bubble anymore, is called flat. The people who make soda add the bubbles to make the soda taste better. The bubbles come from a gas called carbon dioxide. As long as the cap is screwed on the soda bottle tight, or the can is unopened, the carbon dioxide stays inside the soda and keeps it tasting fizzy good.

You need
- 2 identical cans of soda
- use of a freezer
- 2 large drinking glasses

Does freezing soda and then thawing it cause the carbon dioxide gas to leave the soda and make it go flat? Put a can of soda in the freezer. Leave it there for only two hours, long enough for the soda to begin to freeze. (We don't want to really freeze the whole can of soda, because the water in the soda expands when it freezes and could break the can open!) Remember to take the soda out of the freezer after two hours or you may always remember the mess it makes!

Place the cold soda on a table, but not in sunlight. Put an identical can of soda next to it. After four hours, open the two soda cans and pour each one into a large glass. Does one have more fizz and foam than the other? Compare them.

Something more

Can you stir the gas out of a soda? Some people do not like a lot of fizz, because the gas is released later in their stomachs and causes discomfort. Is there something you can put into the soda to get the gas out?

Did you ever drop or shake a can of soda and then open it? The soda shoots out very quickly! After a can of soda is shaken, how long do you have to wait before it is safe to open it without worrying about the soda shooting out?

Project 33

LEFTOVER SALT

Collecting salt from ocean waters

Our oceans are full of salt. If you were out on the ocean in a little boat and were thirsty, you would not be able to drink the ocean water. The salt in it would only make you more thirsty. In some places in the world, oceans dried up long ago and left the salt behind.

Hypothesize that you can show how salt is left when saltwater evaporates. Have an adult help you measure ¼ cup of hot tap water and pour it into a wide-mouth drinking glass. Measure and pour another ¼ cup of hot water into another wide-mouth glass. Into one glass, pour ½ teaspoon of salt. Stir it. Place both glasses in a sunny window. Each day, look at the two glasses and write down what you see. Is something building up on the bottom of the glass of saltwater?

You need

- ½ teaspoon measure
- measuring cup
- an adult
- hot water from a sink
- salt
- 2 wide-mouth drinking glasses
- a sunny window
- pencil and paper

Something more

Use the evaporation process to remove salt from the water of a saltwater lake, river, or ocean nearby. Pour some of the salty water on a cookie sheet. When the water you poured has evaporated, pour more of it onto the sheet. Keep adding more water, as the water on the cookie sheet evaporates. You will soon see salt from the water slowly begin to build up on the sheet.

Project 34

FREEZE OR DON'T?

Preserving with cold

Preserving foods means doing something to make them stay fresh longer. Many foods can be kept fresher for a longer time if they are stored where it is cold. Milk stays fresh much longer if it is stored in a refrigerator than if it is left on the kitchen table.

Some foods are preserved by freezing them. Look in your refrigerator's freezer and see what is being stored in it. You may find ice-cream, meat, and vegetables such as frozen peas, carrots, corn, and lima beans.

What about other kinds of food? Can we use freezing to keep all foods, or just certain kinds? Let's do a test.

Put a fresh piece of carrot in a small, plastic food bag. Put a fresh piece of lettuce in another small plastic bag. Put a fresh piece of celery in a plastic food bag. Place the three bags in a freezer. Leave them there for a day or two.

Remove them from the freezer, and let them thaw out for a few hours. Then look at them carefully; touch them and taste them. Write down any changes you see and any difference in taste caused by the freezing. Is the lettuce still as crisp as it was before it was frozen? Do you think freezing is a good way to keep these vegetables longer?

You need
- small, plastic freezer or storage bags
- carrot
- lettuce
- celery
- use of a freezer
- pencil and paper

Something more
Try to preserve other foods by freezing, such as an egg. Put it into a plastic bag to keep the freezer clean in case it should break. Try to freeze pieces of fruit.

Project 35

GARBAGE HELPERS

Decomposition by tiny animals

Big animals, little animals, and even animals so tiny we can't see them with our naked eyes eat food and help break it down. After your family has eaten a chicken dinner, take two similar leftover bones (such as drumsticks or thigh bones) that still have some chicken on them. Put one leftover chicken bone in a plastic sandwich bag and store it in a freezer. Put the other one on the ground outside of your house. It should be in a place where

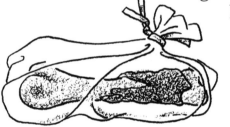

You need
- 2 chicken bones with scraps on them
- colander
- brick
- clothespin
- plastic sandwich bag
- refrigerator freezer

no one will disturb it. Cover the chicken bone by placing a colander upside down on top of it. Put a heavy brick on top of the colander to help keep cats and dogs away from it.

After a few weeks, look at the chicken bone under the colander. Do not touch it. Use a clothespin to hold it up and look at it. Compare it to the one that was kept in the freezer. Which one decomposed, broke down, or rotted faster? Did you see any small animals (ants, flies, etc.) eating from the chicken bone that was left outside?

Something more
Try decomposing other types of food like celery, potatoes, radishes, and a piece of bread. Put some in the freezer, some on the ground, and bury some.

54

Project 36
SMART SPROUTS
Vegetables that start growing by themselves

Plants need moisture and food to grow. Some vegetables, like onions and potatoes, have inside of them everything they need to start growing— all by themselves.

Put two fresh onions and two fresh potatoes in a shoe box. Put the lid tightly on the shoe box so that it is dark inside. Find a cool, out-of-the-way place in your home to put the box, such as your bedroom closet. At the end of one week, open the box. Carefully look at the vegetables. Can you see any changes? Write down what you see.

Put the lid back on the box and wait another week. Then look at the vegetables again. Now what do you see? Write it down. Look at the vegetables every week for four weeks, and write down what happens to them

Something more
A carrot has food inside of it, but often not enough water for the carrot to grow on its own. It needs a little help.

Take a fresh carrot and push four toothpicks partway into it, around the middle. Pour some water in a tall glass. Place the carrot in the glass so that the toothpicks rest on the rim and keep it from falling in. Be sure there is always enough water in the glass to reach the bottom of the carrot.

Project 37

TEA TAMPERING

Changing the pH of tea

Many people like to drink tea. Some like lemon in their tea, others like milk. Does adding milk to tea make it less likely to upset your stomach? It might, if adding milk changes the amount of acid in tea and makes it milder.

The amount of acid in a liquid is measured by "pH". The pH scale goes from 1 to 9: the lower the number, the stronger the acid.

First, pour a glass of water from your sink. Use litmus paper to find out the pH of your water. Write it down.

Then have an adult help you make a cup of tea. Use litmus paper to find out how much acid (called tannic acid) is in the tea. Write the amount down.

Add milk to the tea and stir. Now check the pH of the tea again. Does adding milk raise the pH and make the tea less acidy? If the pH of the tea is raised, do you think the tea would be less likely to upset your stomach? What is the pH of milk?

You need
- a glass of water
- litmus paper with a pH scale of 3 to 8
- an adult
- a cup of tea
- milk

Something more
What is the pH of tea with lemon? What is the pH of different kinds of soda?

Project 38

ROCK AND ROLL-OVER

Looking at life under rocks

What kinds of living things make their homes under rocks? Find out by turning over several large rocks in your neighborhood. You may find ants, larvae, worms, millipedes (many legs), centipedes (a hundred legs), spiders, crickets, or beetles; but watch out for creatures that bite or sting.

Draw pictures of the animals you find living under the rocks. Describe their body parts, the number of legs they have, their coloring, and other things about them. Get a book on insects from the library and try to identify the animals you saw. If some of the rocks you looked under were in a wet place, such as near a lake, and some in a dry place, did you find different kinds of animals under each, or the same kinds?

Something more
Are the animals that you find under a rock the same kind as those you find under an old board or log?

You need
- several large rocks in your neighborhood
- a strong stick
- an adult
- pencil and paper

Camel Cricket
6 legs
dry
brown

Carrion Beetle
6 legs
damp
black

Slug
no legs
wet
grey

Scorpion
stings!

Worm
no legs
damp
pinkish

Project 39

TATTOOED TRAVELLERS

Finding out where insects live

You may have lifted an old board somewhere lying on the ground and seen some tiny animals (beetles, worms, etc.) crawling there. Is this their home, or are they just passing by? If it is their home, can they find their way back there if they are moved a short distance away?

Find an old board or a rock nearby that has been sitting on the ground for a long time. Lift it up carefully with the stick. Are there a number of tiny animals living there? Get a small craft paintbrush and a small bottle of white paint, the kind used for painting models. Use the brush to put one small drop of white paint on the back of any animals you see under the board. Put the board back in place.

The next day, look under the board again. Are the animals with the white markings still there? If they are, pick them up and place them on the ground a step or two away from the board. Put the board back. If you can't stay to watch, go back and lift up the board later on to see if they found their way back home.

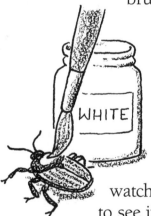

You need
- small paintbrush, from a craft or toy paint set
- white paint, the kind used for modelling
- an adult
- a board that has been on the ground for a long time
- a strong stick
- rubber gloves

Something more
Can you identify any of the kinds of animals that live under the board? Get a book on insects from the library.

Project 40
SALT-FREE
Water purification at home

Most of the oceans and seas in the world are made up of saltwater. When the water evaporates, salt is left behind so the ocean becomes even saltier. But is evaporated water still salty? Let's evaporate some saltwater, then bring the water back to see if it is salty, or if the salt is gone.

Ask an adult for a small clean can (the inside edge can be sharp where the lid was cut away, so you need to be careful). Fill the can about one-quarter full of salt. Add water to it until the can is about three-quarters full. Stir it slowly until the salt is completely dissolved.

Take a large, wide-mouth jar: a mayonnaise or peanut butter jar works well. Clean the jar thoroughly; then, without spilling any of the saltwater, lower the can to the bottom of the jar. Screw the lid onto the jar tightly; this makes it a "closed environment."

Put the jar in a warm, sunny window for several hours. The sun's heat will cause water to evaporate. When the sun starts to go down, carefully place the still-closed jar in the refrigerator. The change in temperature from hot to cold will now make the evaporated moisture in the glass jar condense, and water droplets will collect on the inside of the jar.

Open the jar, reach in, and taste the condensed droplets of water that have formed on the inside of it. Do you taste any salt, or does the water taste clean and pure?

You need
- an adult
- smallest-size, flat (tuna) can
- salt
- water
- spoon
- large, wide-mouth jar with lid
- a sunny window
- use of a refrigerator

Something more
Can evaporation purify water, even if it is dirty instead of salty? Repeat the experiment above, but instead of mixing salt and water in the small can, mix a little soil into the water.

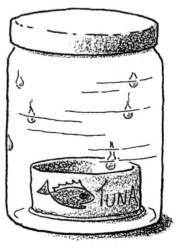

59

Project 41

TREES WITH RAINCOATS

Proving that tree leaves are waterproof

Trees get the water that they need to live through their roots deep in the ground. Even though rain falls on the leaves, trees do not get water through them. In fact, the tops of the leaves are waterproof. This also keeps the water already in the tree from evaporating. Can we prove that tree leaves are waterproof?

Stack some books on top of each other so that the pile is two to three inches (5–8 cm) high. Lay two rulers against the books to make a ramp. The rulers should be spaced three to four inches (8–10 cm) apart. Place a piece of paper towel underneath the ramp.

Gather about ten large tree leaves: broad ones like the leaves from sycamore or some oak trees. Lay the leaves down on the rulers, starting at the bottom of the ramp and overlapping the leaves as you go higher. The tops of the leaves on your ramp must face upward and their bottoms down. When you have completely covered the entire

> **You need**
> - several thick books
> - 10 large, broad tree leaves
> - 2 rulers
> - paper towel
> - small glass of water
> - a pencil or stick

ramp with the overlapping leaves, take a pencil or stick. Press it down between the rulers from the top to the bottom to make a little trough, or path, for the water to follow when you start to pour.

Place a few sheets of paper towel at the bottom of the ramp to catch any water that runs down and off the leaves. Slowly, start pouring a little bit of water at the top of the ramp. Pour it in the middle, between the rulers, so that the trough will keep the water from spilling over the sides. If the leaves are waterproof, the water will run down the leaves and the paper towel placed underneath the ramp will stay dry.

What happened? Are the leaves waterproof?

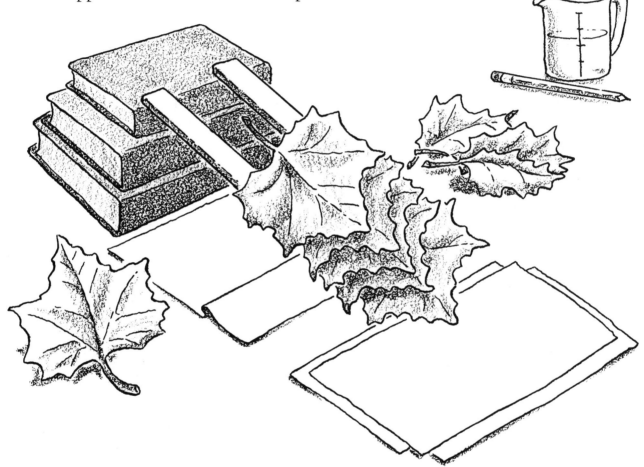

Something more
A tree's leaves act like an umbrella. After a rain shower, look at or feel the ground under a tree and out in the open. Is it drier under the tree? That means it could be a good place to stand to stay dry during a rain shower. But never stand under a tree during a thunderstorm. It is a very dangerous thing to do, because the tree could be hit by lightning.

Look at the bottoms of the leaves. Are they dry? Do pine needles float? If so, they may be waterproof.

Project 42

STILL POLLUTED!

Discovering invisible pollution

We use so much water every day that we often forget just how important it is to us. Make a list of all the things water is used for around your house. Without water, we would not be able to live.

Because water is so important, we must keep our sources of water clean.

Once water is polluted, or dirty, it is often hard to make it clean again. Sprinkle about ten shakes of pepper from a pepper shaker into a small glass of water. Let it sit for two or three minutes. Put a paper coffee filter over the top of an empty drinking glass. Slowly, pour the pepper-polluted water through the filter and into the glass. Did the filter remove all of the black pieces of pepper? Does the water look clear again? Even if the water *looks* clear, maybe the filter did not really get rid of all the pepper pollution.

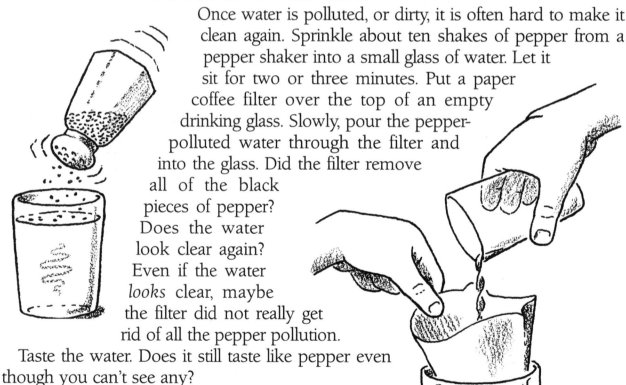

Taste the water. Does it still taste like pepper even though you can't see any?

Something more
Pollute clean water with other types of foods, like salt, and then use a filter to try to make the water clean again.

Project 43
SMELL POLLUTION
When unwanted odors stay in the air

Air pollution can mean unpleasant smells in the air. Some foods have unpleasant smells.

Have an adult cut a slice of fresh onion and place it on a dinner plate. Hold it up to your nose. Is the strong smell unpleasant? Does it make your eyes water?

Pour a capful of lemon juice over the slice of onion, completely covering it. Hold the slice up to your nose again. Is the onion smell as strong? Do your eyes still water?

If lemon juice takes away the unpleasant odor of onion, do you think it would be a good idea to add lemon juice to recipes that have onions in them, like tuna-fish salad?

You need
- fresh onion
- a dinner plate
- lemon juice
- an adult with a kitchen knife

Something more
Do other citrus fruit juices also take away strong food smells? Try pouring orange juice on a piece of fresh garlic.

If you eat a piece of garlic, will drinking orange juice make your breath smell better?

Project 44
PLANT STRAWS
These roots are made for drinking

Some plants and trees have many little roots. Others have one larger, long root going deep into the ground. But add up the "sucking" area of the small roots, with all their tiny parts, called "root hairs," and you can understand how this root gathers more water faster than a larger root: its "surface area" is so much bigger. When it rains, the hairy root can quickly take in the water the plant or tree needs to live, before the rain soaks too far down into the soil. Having more root surface area makes it easier for plants to catch available water.

<table>
<tr><td>You need
• a piece of cotton clothesline
• 2 small, baby-food jars
• water</td></tr>
</table>

Discover how much more water a lot of small roots can gather, compared to one large root. Take one end of a piece of cotton clothesline. (Have an adult cut a piece for you.) Unravel and spread out about two inches (5 cm) of the clothesline, until all of the threads are separated. Fill two small jars half-full of water. (Be sure the jars are exactly alike and filled with the same amount of water.) Dip the end of the clothesline with the loose strands into one of the jars. Let the strands just touch the bottom of the jar. Hold it there and count to ten. Then pull the clothesline out.

Take the other end of the clothesline, with the strands still wrapped tightly together. Dip that end into the other jar until it just touches the bottom. Hold it there and count to ten. Then pull the clothesline out.

Compare how much water is left in each of the jars. Did the end of the clothesline with more individual "roots" gather more water than the "single root" end?

Something more
Look at the root systems of different kinds of plants or vegetables. Identify the plants and write down which kind of root systems they have.

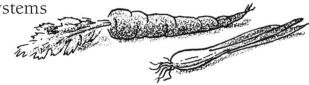

Project 45

RUNNING HOT AND COLD

Testing air and soil as insulators

Many living things find shelter in the ground from hot and cold temperatures. Some animals sleep in the ground during the cold winter months. The ground protects plant bulbs and root systems from cold temperatures.

Fill two small soup cans with warm water from a sink. Put a thermometer in each can. Place one of the small cans in the middle of a bigger can. Fill the space in between the two cans with soil.

Place the other small can inside a bigger can, too, but don't put anything in between the two cans for insulation.

Write down the temperature showing on the thermometers. Every five minutes, look at the thermometers and write down the temperatures. Do this for a half hour.

Make a graph showing the rate of temperature change. Did the can of water surrounded by soil keep its heat longer than the one surrounded by air? Is soil a better insulator than air?

Something more
Is the best heat insulator also the best cooling insulator? Try other kinds of insulation, such as water or leaves.

You need
- 2 small soup cans
- 2 large cans about 5 inches (13 cm) in diameter
- 2 thermometers
- soil from your yard
- warm water
- a watch or clock
- pencil and paper

Project 46

HOLE IN THE SOIL

Comparing ground temperatures under trees

Deciduous trees are trees that lose all their leaves in the fall. An oak tree is one kind of deciduous tree. Conifers are trees that stay green all year. A pine tree is one kind of conifer. During the winter, is the ground underneath a pine tree warmer than the ground underneath an oak tree?

Find a pine tree and an oak tree in your yard, at your school, or in a neighborhood park.

Underneath each tree, about two feet (60 cm) from the trunk, push a pencil or narrow stick into the ground to make a hole several inches deep. Put a thermometer into each hole. Wait a few minutes before reading the temperatures. What is the temperature under the pine tree? What is the temperature in the ground underneath the oak tree? What is the air temperature? Does it make a difference if the day is sunny or cloudy? What does the ground look like around each tree? Are there leaves piled around the oak tree?

Something more

Do the needles of a pine tree make a better blanket against cold than leaves? How would you go about setting up a project to test this guess, or hypothesis?

Are fruit trees any different?

Project 47

SCREENED INSECTS

Collecting and identifying neighborhood insects

What kinds of insects live in your neighborhood? Collect insects from windowsills around your home. An undisturbed screened window in an attic, garage, or shed, where insects are often trapped, may have a lot of dead insects gathered there. (Remember, spiders are arachnids, not insects. Insects have three pairs of legs, but arachnids have eight legs.)

> **You need**
> - dead insects
> - a book on insects
> - pencil and paper

Collect as many insects as you can find and identify them. You may want to mount the insects on a display board, and provide labels giving their name and something about them.

Quantify the number and types of insects you found. In other words, how many different kinds of dead insects did you find on the windowsills? Were there many more of one kind of insect than another?

Something more
Were any of the insects trapped on the windowsills still alive? Are the living insects helpful or harmful?

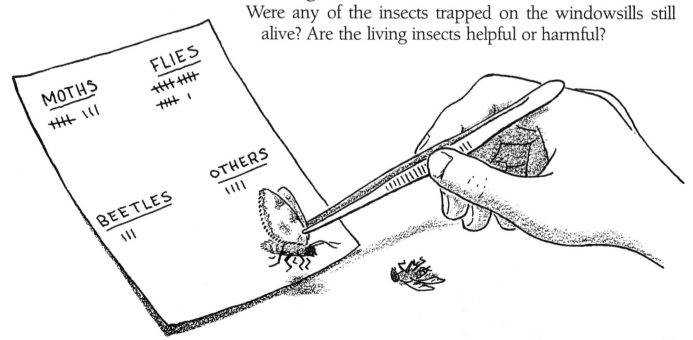

Project 48

GET A GRIP

Comparing the wraps that keep food fresh

You need
- 4 brands of plastic food-wrap
- a ruler
- masking tape
- spring-type clothespins
- scissors
- a table
- pencil and paper

We have many ways to preserve food, that is, to keep it fresh. After dinner, when there are leftovers, people often wrap them in plastic wrap, or put the food in bowls and cover them with the wrap.

If a plastic wrap is going to keep air out so that it will keep the food fresh, it must be able to cling well when it is wrapped around food (clings to itself) or over the top of a bowl (clings to bowl). Is there a "cling" difference in the brands of plastic food-wrap? Do some wraps cling better than others? Do a test and use science to help you be a better consumer.

Cut two pieces of a brand of plastic food-wrap. Each piece should measure one foot (30 cm) long by one foot wide. Use masking tape to attach one piece of the wrap to the edge of a table, so that it hangs down. Take the second piece of wrap and press it onto the bottom of the hanging piece, overlapping the pieces by two inches (5 cm).

Hang a spring-type clothespin on the bottom piece of wrap. Keep hanging more clothespins until the bottom piece of wrap falls off. Write down the number of clothespins the wrap held before it lost its grip.

Do this again for other brands of plastic wrap; include a store or generic (no-name) brand. Which food-wrap tested had the most cling? Which one had the least?

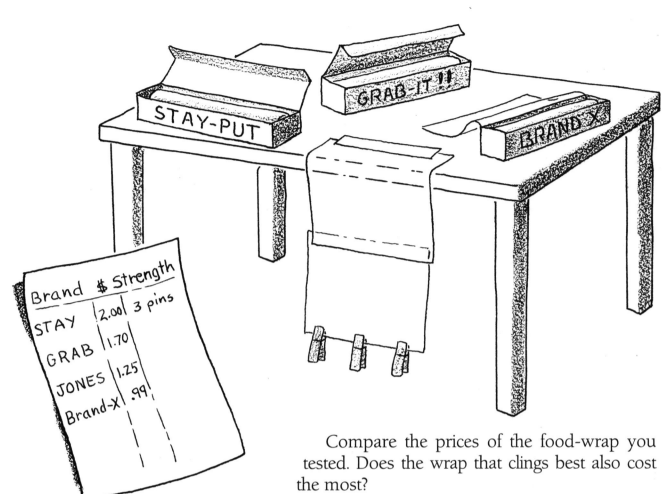

Brand	$	Strength
STAY	2.00	3 pins
GRAB	1.70	
JONES	1.25	
Brand-X	.99	

Compare the prices of the food-wrap you tested. Does the wrap that clings best also cost the most?

Which brands of wrap are the easiest and the hardest to get off the roll?

Something more

What about bowl cling? Put some coins or small stones into a bowl and cover it with the plastic wrap. Does it hold when you turn the bowl upside down? Watch out! Does the kind of bowl make a difference; that is, if it is made of plastic or glass? Does it matter if the outside of the bowl is wet or dry?

Do all brands of plastic wrap do a good job of keeping foods fresh? Cut five slices of apple. Wrap four of them in different brands of wrap. Mark them so you know which wrap is which. Leave one slice of apple unwrapped. Put them all in the refrigerator. Look at the slices every day. What do the apple slices look like at the end of a week?

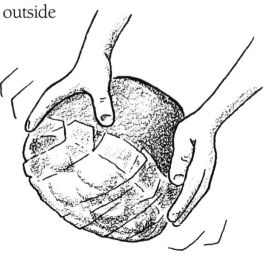

Project 49

A SWALE DAY

Earth's grassy armor

You need
- large pitcher
- four cookie trays, about 18 inches (45 cm) wide
- soil
- a piece of soil with grass growing in it to fit in tray
- kitchen measuring cup
- water
- several thick books
- garden hand shovel

A swale is a low area covered with grass or plant life that water runs through. The plant life keeps the soil from eroding. Fast-moving water can cause very damaging erosion, but the roots of grasses and plants help keep soil from being washed away. Prove that the roots of plants can help stop soil erosion.

Pour three measuring cups of water into a large pitcher. Fill a cookie tray with soil. There should be enough soil in the cookie tray so that it is as high as the tray's lip. Raise one end of the tray by piling several books under it. The raised end should be about 4 inches (10 cm) high. Put another tray under the low end (the bottom of the ramp). This will catch the water that runs down the raised tray ramp.

Put three cups of water into a large pitcher. Pour the water out of the pitcher over the raised end of the tray. Tilt the pitcher so that the water pours out of it very quickly. The tray at the bottom will catch the water and also any soil that the water has eroded away. Put these trays aside for now.

Get permission to take a piece of soil with grass growing on it from your yard or from an area in the neighborhood. Use a small garden hand shovel to cut a piece of soil large enough to fit a cookie tray. The piece should be about 2 inches (5 cm) deep to be sure to get the grass roots. Put the soil piece into a cookie tray, being careful to keep it all together in one piece. Put books under one end of the tray to raise it to the same height as the other raised tray. At the low end, slide another empty cookie tray under it. Pour three cups of water into the 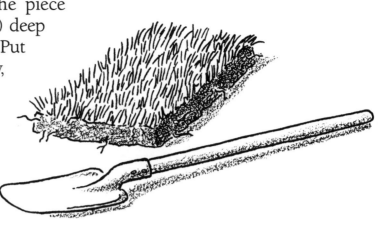 pitcher. Pour the water out over the raised end of the tray. Tilt the pitcher the same as you did before, so that the water comes out very quickly. The tray at the bottom will catch the water and any eroded soil.

Look at the two trays that collected the water that ran off. Does one have more soil particles in it than the other one? Which one has more? Did the piece of soil with grass growing in it keep more of its soil than the other one?

As soon as the experiment is finished, put the piece of grass soil back in the ground where you found it so that the roots will catch and the grass will continue to grow.

Something more

Evaporate the water out of the two collecting trays. Weigh the soil that is left in each one and write down their weights.

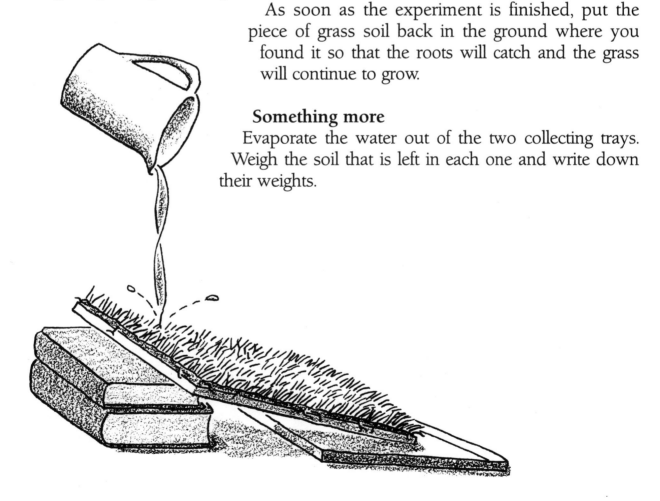

Project 50

MUD HUTS

How to make a heat-retaining adobe brick

Many people live in houses made of wood. In the southwestern United States and in Mexico, where there are not many trees, people often build houses out of mud bricks called adobe. Even the ancient Babylonians and Egyptians, who lived in treeless desert areas, used adobe bricks. These bricks were made from sandy clay, water, and straw. They would bake the bricks in the sun for several weeks, and then use the dried, "cured" bricks to build their houses.

Adobe houses are warm in the evening and cool in the daytime. If a mud brick is warmed by the sun, how long will it continue to give off warmth once the sun goes down?

You need
- soil
- water
- bowl
- large mixing spoon
- straw or dry grass
- 2 thermometers
- one-pint milk carton
- clock
- a sunny window
- pencil and paper

Gather some straw. (If you do not have straw you can use dry grass, or dry needles from under a pine tree.)

Put the straw, soil from your yard, and water into a bowl and mix it well. Open the top of an empty one-pint milk carton. Pour the mud mixture from the bowl into the milk carton. Make a hole in the mud by pushing a pencil halfway down in the middle of the opening. Loosen the mud around the pencil by moving the pencil in a small circle, then leave it in the carton. Put the milk carton in a sunny window and leave it there for several days to dry.

When the brick is firm and dry, take the pencil out of it and peel off the carton. Leave your brick in a sunny window for one more hour. Then, put the brick on a table out of the sunlight.

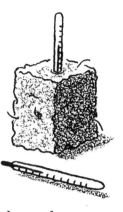

Put a thermometer into the hole in the brick. This will measure the temperature inside the brick. Lay another thermometer nearby on the table to measure the temperature of the air outside the brick. Will the thermometer inside the brick measure a hotter temperature than the one outside the brick? Wait a few minutes, then read and write down the temperatures showing on the thermometer inside and outside of the brick. How long will it take before the thermometer inside the brick is the same temperature as the one outside of it?

Something more

Adobe bricks are not used for building in places where there is a lot of rain, or where it is cold. What do you think would happen if adobe bricks froze and thawed a lot? What happens to adobe bricks if they keep getting wet?

What could be added to the mud mix to make stronger bricks? Build a scale-model town out of adobe.

Project 51

A HOT LUNCH

Pets choose between heated or cooled foods

People eat foods at different temperatures. We eat cold ice-cream, room-temperature peanut-butter-and-jelly sandwiches, and hot hamburgers. Pet cats and dogs usually eat their food at room temperature. Would they eat their food if it was cold? Would they eat it if it was hot? Do they like their food more if it is heated or cooled? What do you think your pet will like?

Put a small amount of pet food in three bowls. Cover each bowl with plastic food-wrap. Put one bowl of food in a warm sunny window. Put another bowl in the refrigerator. Put the third bowl on a table that is out of direct sunlight.

After about an hour, take the plastic wrap off the bowls and put all three down where your pet usually goes for food. Watch your pet. What is its reaction to the three bowls. From which bowl does your pet eat? Does it eat from more than one bowl?

> **You need**
> - a pet cat or dog
> - pet food
> - 3 feeding bowls
> - plastic food-wrap
> - a sunny window
> - refrigerator

Something more

Does heat produce a stronger food smell? If so, maybe it is the stronger smell from the warm food that attracts your pet, rather than the temperature of the food. How could you find out if this is true?

Do cats and dogs react the same to different-temperature foods? Do some cats and some dogs react differently, really prefer hot or cold food?

Project 52

TRAPPED RAYS

Plastic food-wrap as an insulator

Is plastic wrap good to keep in warmth? Fill two small, plastic butter or margarine tubs with soil. Add water to the soil and mix to make mud. Push a thermometer into the middle of each tub of mud. Set both tubs in a warm, sunny window for one hour.

An hour later, take the tubs out of the sunlight. Read the temperature showing on each thermometer, write it down (the temperatures should be the same), and put the thermometer back. Cover one of the tubs with some plastic wrap, wrapping it tightly around the thermometer. With the pencil, make a small hole (for the thermometer) through two or three more sheets of wrap. Place these over the wrapped tub, too. Be sure the plastic film is wrapped tightly around the whole tub.

Every fifteen minutes, look at the thermometers and record their temperatures. How long does it take for the two tubs of mud to lose their heat and reach room temperature? Did the tub covered with plastic wrap keep its warmth longer?

You need
• 2 small, plastic butter or margarine tubs
• plastic food-wrap
• 2 thermometers
• soil
• water
• spoon
• sunny window
• pencil and paper

Something more

Do the experiment again, but this time cover one tub with one sheet of plastic wrap, and cover the other tub with two sheets. Does the tub wrapped with two sheets of wrap keep its warmth longer?

Test the ability of other things to keep heat in, like aluminum foil or paper napkins. What is the best insulator you can find?

Project 53

TREES, PLEASE

Wind protection for your home

Trees seem friendly. They are colorful and nice to look at. People hang swings from them. Birds build nests in them.

In places where the winters are cold, strong winds can make it harder to heat a house. Do trees stop the wind? If trees are planted around a house, can they help make it easier to heat the house?

On a windy day, go with an adult to an open place away from buildings and trees. Put a stone or small rock on the ground.

Take a plastic-foam drinking cup and, with your fingers, break off four small pieces. Hold one piece in your hand. Raise it in front of you as high as you can reach above the rock on the ground. Be sure you are standing

You need
- a place with many trees
- an open place, away from trees and buildings
- an adult
- plastic-foam drinking cup
- measuring stick or steel rule
- a windy day
- stone or small rock
- pencil and paper

so that your body is not blocking the wind. Let go of the piece. Mark the spot where it first hits the ground. Use a measuring stick or steel ruler to see how far the wind blew the plastic-foam piece away from the rock. Write down the distance. If it is a very windy day, the piece of the cup may blow too far. If that happens, use something a little heavier to drop, like a small piece of cloth. The cloth can be wet a little to make it even heavier, if needed.

Do this test again three more times. Each time, use a different piece of plastic foam or cloth to drop. Measure and write down the distances from the rock to the piece.

Add the four distances together, then divide the answer by four. This gives you the average distance the wind blew the pieces.

Pick up all the plastic-foam or cloth pieces and the rock when you leave. Do not litter.

Next, find a group of trees. The trees must have branches that go all the way to the ground, or there must be many bushes or shrubs surrounding the trees. Put the rock on the ground close to the trees. Again, standing by the rock, raise your hand in front of you as high as you can reach and let go of the piece of plastic foam. Measure how far away the wind blows the piece from the rock. Write down the distance.

Do this again three more times. Use a different piece of the foam cup each time. Measure and write down the distances from the rock to the piece of cup.

Add the four distances together. Divide the answer by four to get the average distance the wind blew the pieces.

The stronger the wind, the farther away the pieces of plastic foam will be carried. Which average distance was larger, the pieces dropped in the open space, or near the trees? What does this mean?

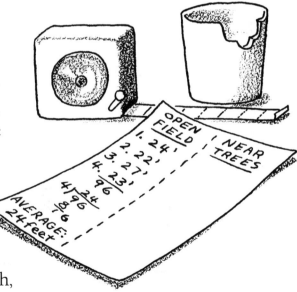

Something more
On which side of your house should you plant trees and shrubs to help protect your home from the winter winds (north, south, east, west)?

On which side of your house should you plant trees and shrubs to help protect it from the hot summer sun (north, south, east, west)?

Why do you think farmers might want to use trees to stop the wind?

Project 54

"CHEEPER" FOOD

Do birds prefer popped or unpopped corn?

Birds like to eat seeds. Popped popcorn is a delicious treat for people, but corn is not found that way in nature. Bread is not found in nature either, but birds like it. What do you think birds would like best to eat, unpopped popcorn or popped popcorn?

> **You need**
> - 2 small, plastic margarine or butter tubs
> - stiff piece of cardboard
> - glue
> - popcorn
> - a window

Glue two small, plastic margarine or butter tubs near one edge of a stiff piece of cardboard The cardboard should be about the size of a sheet of typing paper. Fill one tub with unpopped popcorn and the other tub with popped popcorn. The popcorn should be plain, without butter.

Put the cardboard on a windowsill with the tubs on the outside. To hold it and keep the wind from blowing it away, close the window on the cardboard. Now watch to see if any birds come to eat at the windowsill. Do they eat the popped corn, or the unpopped corn?

If no birds come to look at the popcorn, put a little birdseed or bread on the cardboard. This will be the bait, to attract them to your windowsill, then wait to see if they eat any of the popcorn from the tubs.

Something more
Identify the kinds of birds that come to your window. Pick up a book on birds at the library. Do some species of birds prefer the popped corn over the unpopped corn?

Try other kinds of "people food" to see if birds like them or not. You can serve the birds breakfast cereals, fruits, vegetables, and liquids (soda, milk, juices) in the tubs.

Project 55

RIME OR REASON

Making water vapor visible

Did you ever go outside early in the morning and find the ground wet, even though it had not rained? That water all over the ground is dew.

In the air are tiny droplets of water. Most of the time these water droplets are hidden. They are too tiny to be seen. If there is a lot of water in the air, we see it as fog. But even when there is not enough water to make fog, some droplets are there. When the temperature of the air rises to a certain point, the water in the air cannot stay there. The temperature at which the water comes out of the air is called the dew point.

Is there water in the air in your room right now? Cold air can make the water vapor visible. Fill a plastic 1-litre soda bottle three-quarters full with water. Screw the cap on and place the bottle in the freezer. Wait until the water has turned to ice.

Take the bottle out of the freezer and set it on a dinner plate on a kitchen counter or table. The ice in the bottle will cool the air that comes near it. As the air cools, the hidden moisture in the air will appear on the outside of the bottle. Is white frost instead of water droplets forming on the bottle? That happens if the bottle is very cold. The white frost is called rime. It sometimes appears on window panes during very cold winter nights.

You need
- 1-litre plastic soda bottle
- dinner plate
- kitchen table
- water
- use of a freezer
- pencil and paper

Something more

If the air coming out of your lungs is much warmer than the air around you, a cloud of moisture may be seen. Breathe on an eyeglass lens. What happens to the eyeglasses? What happens if the glasses are warmed first?

Project 56

MOISTURE, MOISTURE, EVERYWHERE

Testing humidity indoors and out

In the air all around us there are tiny droplets of moisture. We call this moisture that is in the air humidity. If there is a lot of moisture in the air we say that the humidity is high. When someone takes a hot shower, the room steams up. We can easily see that the humidity there in the bathroom is high.

Would you guess that the humidity in your home today is higher or lower than it is outside?

Cut two squares out of an old piece of cloth. Each piece should measure four inches (10 cm) square. Fold each piece in half, and then in half again.

You need
- 2 rulers
- 2 pieces of cloth, 4 inches (10 cm) square
- 6 thumbtacks
- 2 pencils
- safety scissors
- eyedropper
- paper
- clock or watch

Once you finish preparing the cloth, lay a pencil on a table. Set a ruler across the pencil, at about the middle so that it looks like a see-saw. Place one folded piece of cloth at one end of the ruler. Lay one thumbtack upside down on the cloth. Lay a second thumbtack upside down near the other end of the ruler. Move the pencil slowly under the ruler until it balances as well as possible.

Now, put another thumbtack upside down near the end of the ruler that does *not have* the piece of cloth on it; this will weight down that end. We now know that when the cloth is dry, the other end of the ruler will be heavier.

Fill an eyedropper with water. Slowly squeeze drops of water onto the cloth until the ruler tips and the water-soaked cloth side is now the heavier side. Be sure to count the number of drops you squeeze. Look at a clock or watch and write down the time. Keep

checking the ruler balance. When the side of the ruler with the two thumbtacks touches the table, write down the time again. How long did it take for most of the moisture to dry out of the cloth?

Make another seesaw ruler, just like this one, outside. Squeeze the same number of water drops onto the cloth as you did inside. Write down the time and keep checking the balance. When the side of the ruler with the two thumbtacks touches the table, write down the time. If the air inside is dryer than the air outside, the water on the cloth in the house will evaporate and the cloth will dry quicker than the water on the cloth outside.

Did you guess right?

Something more

Is there any difference in the humidity of the air outside on a sunny day compared to on a cloudy day?

Is there any difference in the humidity of the air during the day compared to at night?

81

Project 57

UPDRAFT

Temperature and water convection

When a body of air becomes warmer than the other air around it, that body of air rises. It rises because it is lighter and less dense than the other air. Warm air rises and cold air falls. When air moves this way, it is called convection. On a hot summer day, you can see warm air rising from a blacktop road or from a barbecue grill.

Convection happens with water, too. When you go swimming at a lake, you may notice that the water by your feet, when you are standing, feels colder than the water at your chest. During the day, the water in the lake or pond is warmed by the sun hitting the surface. During the night, the water on the surface cools off first. As it does, the water drops down. The warmer water underneath is then pushed up, because cold water is heavier and more dense than warm water.

To prove that this hypothesis is true, pour water into a clear drinking glass. Put the glass of water in the refrigerator. Also, put in the refrigerator a bottle of blue food coloring. Wait one hour, and then take them both out of the refrigerator

> **You need**
> - 2 clear drinking glasses
> - water
> - bottle of red food coloring
> - bottle of blue food coloring
> - use of a refrigerator
> - spoon

Fill the other clear drinking glass with hot water from a sink faucet. Do not make it too hot: you don't want to burn yourself.

With the bottle of red food coloring at room temperature, slowly squeeze six drops into the glass of cold water. Because the cold water is denser, the red coloring stays on top. Slowly squeeze six drops of red food coloring into the glass of hot water. What happens?

Stir both glasses. Then, squeeze six drops of the cold, blue food coloring into the glass of cold water. What happens? Squeeze six drops of the blue food coloring into the glass of hot water. Does the blue coloring sink?

Something more
Instead of using food coloring in the two glasses of water in this project, can you find natural food coloring to use? Try juice from a can of red beets, a weak tea, or some cranberry juice.

Project 58

BAG RECYCLING

Uses for brown paper grocery bags

We must be careful with the things we have. Conservation is using things wisely and not wastefully. If you want to conserve, buy only things that can be used again and again. When cleaning, use reusable rags instead of paper towels. Use regular dishes and glasses for meals, instead of paper plates and throwaway cups.

In addition to using things more than once, you also conserve by using things for more than one purpose. The paper bags that you get from food stores when they pack your groceries can be reused by making them into coasters (the small mats that go under a drinking glass or cup to protect a tabletop).

You need
- paper grocery bags
- glue
- safety scissors
- a heavy book
- large jar lid
- pencil

Lay a paper bag flat on a table. Place a large jar lid on the bag. Use a pencil to trace the outside circle of the lid onto the bag. The coaster must be bigger around than the drinking glass, so check to see that the size will fit the glasses used in your home.

Cut out the circle with scissors. Use the lid again to trace four more circles on the paper bag. Cut them out. Place some glue around the edges of the circles and stick them together, one on top of the other like a layer cake. Be neat. Do not use too much glue, or it will seep out.

Place a heavy book on top of the stack of paper circles and wait until the glue dries.

Cut enough circles (or squares) to make a set of four or five coasters. Decorate the tops with pictures cut from magazines or comic books. The finished coasters can be given as gifts. You could decorate a set of coasters with cancelled postage stamps and give it to a friend who collects stamps.

How good are your paper coasters at keeping a table dry if a very cold drink, with ice cubes, is placed on them? Are five layers of paper enough to keep the table dry? If not, how many layers are needed?

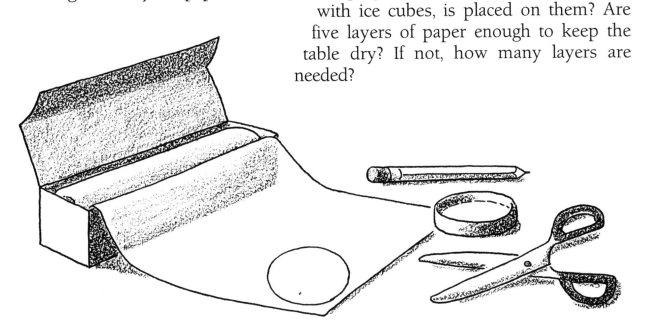

What if you put wax paper or aluminum foil circles in between the paper circles?

Something more

You probably have lots of paper grocery bags. What other uses can you think of for them? Why not use them to make paper dolls, cutouts, puppets, masks, and book covers for school?

When you go grocery shopping, take some paper or plastic bags back with you to pack your groceries at the check-out, so you won't have to use new ones.

Project 59

WHO'S RAPPING, TAPPING?

Determining sound direction

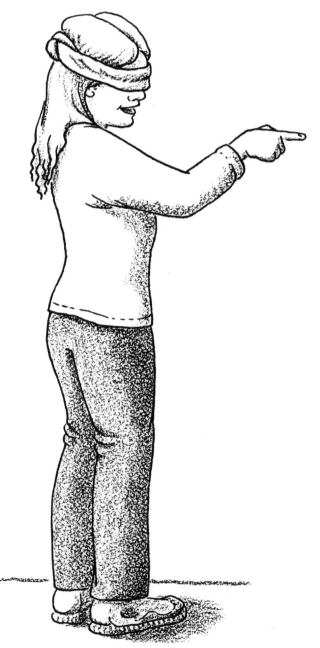

You need
- seven friends
- fourteen pencils
- a blindfold
- a large bowl
- safety scissors
- paper

We have five senses to learn about our environment. Our ears are the wonderful organs that let us hear sounds. Because our ears are placed one on either side of our head (instead of both on the same side), our brain can figure out from the signals where a sound is coming from. Using your two ears, can you find your friends from the tapping sounds they make?

With scissors, cut seven small pieces of paper. Write a number on each piece, from 1 to 7. Put all the pieces of paper in a large bowl. Gather seven friends. Give each person two pencils. Tell your friends to each pick a piece of paper from the bowl. Give one person a sheet of paper. That person will keep count of how many guesses you get right.

Stand in the middle of a room and have a friend blindfold you. Tell your friends to stand in a big circle around you, with you in the middle. One by one, have each friend tap two pencils together three times. The friend who picked the paper marked "1" goes first. After your friend taps the pencils together, point to where you think the sound is coming from. The person who is keeping track of whether your guesses are right or wrong writes it down each time you point.

After the first person has tapped the pencils together and you have made your guess, the friend who has the paper numbered "2" should tap his or her pencils together three times. Again, point to where you think the sound is coming from. Have each friend take a turn as you make your guesses, until all seven have gone.

How many did you get right?

Something more

How far apart in distance do two sounds have to be before you can tell that they are coming from two different places? Have two friends stand close to each other and take turns tapping their pencils. Explain that sometimes the same person should take two turns, to see if you can tell that the sound has not moved.

Try to tell the difference in a sound made about two feet (60 cm) below and above your ear when your friend is standing four feet (120 cm) in back of you?

Cover one ear and then try the tapping experiment with your seven friends again. How many do you think you will get right using only one ear compared to when you use both ears? What if you turn your head between taps?

Project 60

STAR BIG, STAR BRIGHT

Sizing up our sun

The diameter of an object is a measure of how big it is through its center. Even though the sun is very far away, it is so large (its diameter is about 865,000 miles or 1,393,000 kilometres across) that the sun sends us its star light in a wide path. The light even seems to be coming to us from more than one place, because the sun can make more than one shadow of things.

> **You need**
> • ruler
> • sheet of white paper
> • pencil
> • a sunny day

On a sunny day, put a piece of white paper on the ground outside and lay a pencil on it. Then pick up and hold the pencil about a foot (30 cm) above the paper. Do you see two shadows of the pencil? Slowly move the pencil down towards the ground until you see only one shadow. With a ruler, measure how far from the ground the pencil has to be before only one shadow is visible.

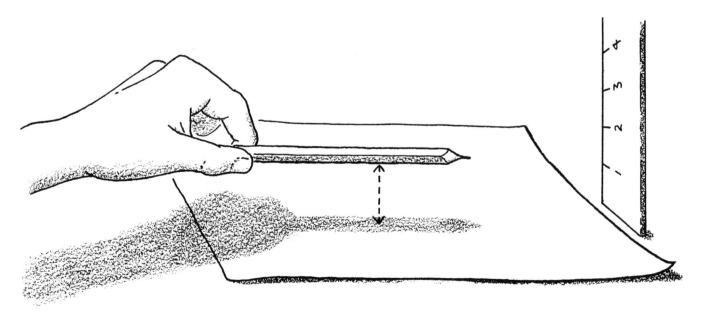

Something more

When air is heated to different temperatures, what we see through it seems to shake. Sometimes, on a hot summer day, if you look far away and close to the ground, you will see things shaking because of this heated air. The same thing happens if you look at something across the top of a hot stove. The object will look as if it is shaking.

Air heated to different temperatures is what makes the stars in the night sky seem to shake, or twinkle. Because the planets are much closer to us than the faraway stars, they look larger and more light from them reaches our eyes, so the planets we see do not twinkle. The light from the planets is not really their own, but is light reflected from the sun. An astronomy book from the library can help you identify what you see in the night sky. At many times during the year you can see the planets Venus and Mars by using only your eyes. Can you see that the stars twinkle and the planets do not?

WORDS TO KNOW
A science glossary

adobe   A mixture of clay, water, and either straw or pine needles made into bricks and baked in the sun. For thousands of years, people living in desert areas, such as the ancient Babylonians and Egyptians, Mexicans, and Indians in the southwestern United States, made their houses out of adobe bricks.

atom   The smallest unit of an element.

chaos   A confused, unorganized condition.

characteristic   Whatever is different or special about something, so that it can be described.

classification   A method of grouping things based on what is the same about them and their characteristics.

compost pile   All kinds of decomposing material that makes soil nutritious to plants.

coniferous tree   A tree that bears cones, such as a pine tree. Trees that stay green all year are called "evergreens." *See also* deciduous tree.

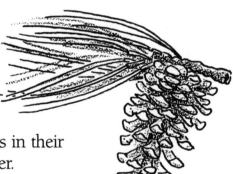

consumers   Those who use up, or consume, things in their environment. It is important to be a wise consumer.

control group   A group that is left alone. The experimental group is later compared to the "left alone" group to see if there are any differences.

convection   The movement of a liquid or a gas caused by temperature. Hot air rises and cold air falls. Cold water falls when placed in hot water.

deciduous tree   A tree that loses its leaves in the fall, as opposed to those trees known as "evergreens." An oak tree is one kind of deciduous tree.

decompose   The breaking down of something into simpler molecules.

density   How tightly packed together the molecules of a substance are. Water is denser than wood, so wood floats in water.

diameter A measurement straight through the center of an object from one edge to the other. The diameter of the sun, for example, measures about 865,000 miles or 393,000 kilometres across.

DIAMETER

element A basic building block from which everything in the universe is made.

environment Everything in the world around us: including air, water, soil, trees, sounds, smells, buildings, furniture.

erosion A wearing away of soil or other matter, usually by wind or water.

evaporation The process that changes a liquid into a gas.

experiment An activity done to prove whether a hypothesis, or guess, is true or false.

germination The point at which a seed begins to grow.

humidity The amount of water vapor there is in the air. Humidity can be high (a lot of moisture) or low (not much moisture).

hypothesis A thoughtful, reasoned guess about something, based on what is known. A hypothesis must be proven by experimentation.

insulator A substance used to keep something at its current temperature, lowering the effect of a different outside temperature.

litmus paper Paper treated to change color depending on the amount of acid or alkaline (measure of pH) in a liquid.

migration Moving from one area to another. Birds and animals usually migrate because of the change of seasons or to look for food.

moisture   Tiny droplets of water vapor in the air or that form on something because of condensation. Also, water in any form.

molecule   The smallest unit of something; much too small to be seen with the naked eye. Molecules are made up of even smaller atoms.

nocturnal   Something more active at night, such as certain animals and plants.

observation   Using your senses— smelling, touching, looking, listening, and tasting—to study something closely, sometimes over a long period of time.

particle   A very tiny bit of something. When two chalk erasers are clapped together, particles of chalk can be seen in the air.

pH   An acid/base scale used to show the alkalinity or acidity of a liquid.

pollution   Anything that is harmful to the environment.

recycle   To use something again and again. Some product packaging uses recycled materials.

rime   Moisture from the air that forms as a white frost, such as on the inside of a window on very cold winter days.

room temperature   The temperature at which most active people feel comfortable, usually about 68 to 70 degrees Fahrenheit (20–21 degrees C).

sample group   A smaller group that takes the place of a much larger group, in order to do a test. The size of the sample group should be big enough to give a true picture of the larger group.

scientific method   A step-by-step way of proving a hypothesis to be true or false.

surface area   The part of an object that is in contact with the outside.

tannic acid   An acid found in tea and in some tree bark.

thermometer   An instrument used to measure warmth.

INDEX

About the Authors

ROBERT BONNET, who holds an M. A. degree in environmental education, has been teaching science at the junior-high-school level in Dennisville, New Jersey, for over twenty years. He was a State Naturalist at Belleplain State Forest in New Jersey and has organized and judged many science fairs at both the local and regional levels. Mr. Bonnet is currently chairman of the Science Curriculum Committee for the Dennisville school system and a science teaching fellow at Rowen College in New Jersey.

DAN KEEN holds an Associate in Science degree, having majored in electronic technology. A computer consultant, he has written many articles for computer magazines and trade journals since 1979. He is also the co-author of several computer-programming books. In 1986 and 1987, he taught computer science at Stockton State College in New Jersey. Mr. Keen's consulting work includes writing software for small businesses, and teaching adult-education classes on computers at several schools.

Together, Mr. Bonnet and Mr. Keen have had many articles and books published on a variety of science topics.